DISASTER RESPONSE
Principles of Preparation and Coordination

DISASTER RESPONSE
Principles of Preparation and Coordination

ERIK AUF DER HEIDE, MD, FACEP
Emergency Physician
Auburn Faith Community Hospital
Auburn, California

with a chapter on the Incident Command System by
Robert L. Irwin
Emergency Management Consultant
U.S. Forest Service, Retired
Sonora, California

The C.V. Mosby Company
ST. LOUIS * BALTIMORE * TORONTO   1989

Executive Editor: Richard A. Weimer
Editorial Project Manager: Lisa G. Cunninghis
Copy Editor: Valerie Gardiner
Book Design: Rhonda Meyers
Indexing: Blue Pencil, Inc.
Cover Photo: Multiple vehicle pile-up on Interstate Highway 5 in Sacramento, California, October 6, 1985. This picture illustrates the type of disaster that is likely to happen in any community. It also represents a major theme in this text; disaster planning should focus on "likely" disaster problems. (Photograph courtesy of Gary Harsh, Sacramento, California.)

Printed in the United States of America

The C.V. Mosby Company
11830 Westline Industrial Drive, St. Louis, Missouri 63146

Library of Congress cataloging in publication data

Auf der Heide, Erik.
    Disaster response: principles of preparation and coordination/
Erik Auf der Heide.
        p.   cm.
    Bibliography: p.
    Includes index.
    ISBN 0-8016-0385-4
    1. Disaster relief—United States—Management.   2. Disaster
medicine—United States.   I. Title.
HV555.U6A94   1989
363.3'48'0973—dc 20
                                                          89-33177
                                                          CIP

CG/MV/MV   9   8   7

To Lee, Laura, and Kristen—who waited

and

to Andrew Di Bartolomeo, MD, FACEP:
residency director, friend, and mentor

FOREWORD

Dr. Auf der Heide has taken on a monumental task. In this book he shows the recurrent problems that exist with the delivery of disaster care. He rightly points out that most failures of disaster management are system problems. I could not agree more.

Over the past twenty years I have been extensively involved in the medical aspects of disaster planning. I am a past chairman of the California Medical Association Disaster Committee and served as chairman of the Office of Emergency Services Disaster Medical Committee. During those years I learned that the public sector is charged, usually by law, to effect disaster planning; yet most of the resources are in the private sector. Seldom is there effective communication between the private and public sector. I also learned that public agencies rarely communicate effectively with each other in regard to disaster planning, compounding errors. It is my strong conviction that there will never be effective disaster medical care until it is incorporated into day-to-day EMS activities. This includes communication, access to transportation, and knowing the hospital resources and how to properly distribute patients. Cooperation between the private sector and the public sector on a daily basis is required before a profound impact on effective disaster response is realized.

Dr. Auf der Heide has addressed these issues in this book and has offered solutions. I believe this is an excellent resource. It should serve to focus our efforts on how to best achieve superior disaster care.

Donald D. Trunkey, M.D.
Chairman, Department of Surgery
Oregon Health Sciences University

PREFACE

Disaster Response: Principles of Preparation and Coordination grew out of a relatively modest project—to create a triage training program for emergency medical technicians—in which I became involved in the spring of 1982. I was then director of the advanced life support ambulance base station at the University of California, Davis Medical Center, and served as a disaster planning committee member at the medical center, for the local medical society, and for the County of Sacramento. Although I was unable to develop the program while at the university, my interest was stimulated. I later carried out an extensive survey of the medical and allied medical literature on triage and on disaster management. There were two things that impressed me in this material.

First, many of the recommendations made in the available literature were not grounded in rigorous, scientifically-based observations. Rather, most of the articles gave advice without producing evidence to show that the advice was correct. In fact, much of the literature did not even examine actual disasters, but only described disaster drills or exercises.

Second, as I read about a number of disasters, it became clear that in too many cases the same mistakes were being repeated by different communities. Sometimes, the *same* community repeated its own errors in successive disasters. This pattern generated concerns about the usefulness of existent disaster literature. Either the suggestions were not very effective, or people were ignoring them. If the advice was being ignored, why was this the case?

As I continued my research, I discovered a body of written material on disasters that existed *outside* the medical literature. There is, in fact, a large collection of research on disasters that has been carried out by sociologists, psychologists, and those involved in the study of public administration and fire science. I found that the most useful material for my purposes resided in the sociological literature.

As I explored this wealth of research, I learned there are often problems affecting the hospital that are the result of actions or omissions by people at the disaster site; this is an area over which the hospital has little control. A common example is the tendency to transport most of the injured victims, frequently by non-ambulance vehicles, to the nearest hospital while other nearby hospitals remain unused. This is but one illustration of how one organization suffers the consequences of actions carried out by another group or individual.

It became apparent that disaster problems cross disciplinary lines. One cannot effectively address disaster management difficulties by focusing on the isolated problems of a single type of organization such as a hospital. Hospitals (as well as many other organizations) are influenced by the activities of a host of other independent agencies including ambulance services, police departments, fire departments, military personnel, the media, etc., whose actions can have a profound impact on their ability to function in a disaster.

In light of these observations, I became convinced that a program to teach triage, by itself, would accomplish very little unless these other more fundamental issues in disaster management were also addressed. I therefore redirected my efforts to deal with the more general problems of disaster response and to take a more interdisciplinary perspective. This perspective is a much needed approach, because, while disaster problems cross disciplinary and organizational boundaries, disaster planning typically does not. On the contrary, the multitude of organizations that may become involved in a disaster response often plan in isolation.

Most disaster response problems are not failures of the individual. More often, they are *systems problems*. That is, the usual organizational systems (procedures, management structures, and designation of responsibilities) established by various organizations to cope with routine, daily emergencies, are not well adapted for use in disasters. Accordingly, this text emphasizes not so much what the individual can do to influence disaster response, but what can be done on an organizational and inter-organizational level. While familiarity with the material in the book will allow emergency medical technicians, firefighters, police officers, or physicians, to see how they fit into the overall picture, it will be of most use to those with organizational management, planning, and policy-making responsibilities.

Addressing the interdisciplinary aspects of disaster response management does not change the fact that emergency medical care is an important focus of this text. This not only reflects my own background and training as an emergency medicine specialist, but also the importance that society places on the numbers of deaths and injuries caused by disasters. Disasters are, in fact, often defined in terms of the numbers of dead and injured.

The material in this text is derived from the research on peacetime, natural, and technological disasters. Disasters of social conflict such as civil disturbances, riots, terrorism, and war are markedly different phenomena, and the conclusions derived in this book may not be applicable in those types of events. Caution also must be taken when comparing the United States to foreign countries where different levels of development, different cultures, and different styles of government may require different approaches. For the most part, therefore, the studies referred to in this book are those that have been carried out in the United States. Applying these studies to disasters that involve tens and hundreds of thousands of casualties must also be interpreted with caution. Peacetime disasters of such magnitude from which we might draw conclusions have simply not occurred in the United States.

The writing style: Although examples of how to manage specific
disaster problems are given, the text is not intended to be a manual, "cookbook," or "how-to" book on disaster management. It is intended to be a basic principles-oriented text. The emphasis is not so much on how, as it is on why.

This book was written with the intention of being an authoritative and well-documented work. Extensive references have been provided to show the sources of data on which conclusions may be derived and principles based. The thorough referencing will also be of help to those who wish to learn more about specific topics. While I have located a large number of useful studies on disasters, many have been difficult to find. Some are unpublished or out-of-print, while others are from diverse and rather esoteric journals, books, and other publications. When possible, addresses are provided. Material that is no longer in print can often be obtained from your local library.

The reader will notice that a number of the facts and conclusions that appear in this book seem contrary to the "conventional wisdom" about disasters. This is because many traditional beliefs have been disproved when subjected to careful examination and well-designed study. It may be that others will challenge some of these conclusions, offer supplemental observations, or derive alternative interpretations. To the extent that this can be successfully accomplished, it will only improve our understanding, and is, therefore, welcomed.

Examples are used extensively in this text, many from actual disasters, to illustrate what happens in these events and how the problems have been successfully or ineffectively approached. Where "conventional wisdom" has been refuted or contradicted, I have tried to provide multiple examples and documentation to support my contentions.

As I live in California, it is natural that I am most familiar with the way things are done there. To the extent possible, I have tried to use examples from other parts of the country and to avoid a west coast bias. Nonetheless, any tendency toward California examples is not intended to suggest that California is any better or worse than other parts of the country, but merely reflects my predominant geographical exposure. In particular, some have expressed the opinion that my emphasis on the Incident Command System (Chapter 7) reflects such a bias. However, to my knowledge, although other management systems exist, none has been accepted on a national basis to the extent that the Incident Command System has.

There is an increasing sensitivity in recent times toward the use of words that express gender. Unfortunately, the English language has not kept pace by providing an array of non-gender-specific third person, singular pronouns. Rather than taking the awkward stance of using he/she, etc., or avoiding the use of any pronouns expressing gender at all, I have chosen to use "he," "his," and "him" in the generic sense. Please understand that I am referring to both the masculine and the feminine situation.

ACKNOWLEDGMENTS

Publishing text with an interdisciplinary approach is faced with obstacles. The same disciplinary boundaries that isolate disaster planning also influence the ways in which textbooks are marketed. While one company publishes professional books for community planners, another for firefighters, another for law enforcement, and another for medical personnel, this book addresses problems facing all of these audiences. It is indeed an act of marketing courage for the C.V. Mosby Company and senior editor Richard Weimer to see this important need and attempt to fill it, and I am grateful for their support.

Over the 6 to 7 years during which this text developed, there were many who assisted me by reviewing manuscript drafts, providing ideas, and obtaining material and information for the project. In particular I would like to acknowledge the following persons:

For extensive reviewing and editing of the later drafts, I owe special gratitude to my wife, Lee.

For reviewing the later manuscript drafts I am also indebted to:

Ralph and Lisl Auf der Heide, Santa Barbara, California

Roeslein Birdsell, Malibu, California

Gene S. Bach, Ranger-in-Charge, Battalion, Chief Tim Lewis, and Administrative Officer Bill Murdock, State of California, Department of Forestry, Santa Rosa, California

Norman Dinerman, M.D., Associate Director of Emergency Medical Services, Denver General Hospital, Denver, Colorado

Thomas E. Drabek, Professor of Sociology, Disaster Researcher, University of Denver, Colorado

Timothy Exline, Sonoma County Fire Services, Santa Rosa, California

Capt. Greg Jilka, Twin Hills Fire Protection District, Sebastopol, California

Harry Martin, Disaster Planner, Emergency Medical Services Agency, County of Sonoma, California

Ted Hard, MD, Director, Emergency Department, Santa Rosa Community Hospital, Santa Rosa, California

Deputy Chief Jerome Ringhofer, Desert and Mountain Command, San Bernardino County Sheriff's Department, San Bernardino, California.

Capt. Bob Waldron, Santa Rosa Fire Department, California

For extensive manuscript review and authorship of the chapter on the Incident Command System:

Bob Irwin, of Basic Intergovernmental Services, Sonora, California, formerly FIRESCOPE Program Manager, and veteran of the U.S. Forest Service.

For providing helpful disaster-related information and material:

Patricia Kuhns, Learning Resource Center Librarian, National Emergency Training Center, Federal Emergency Management Agency, Emmitsburg, Maryland

Dave Morton, Librarian, Natural Hazards Information Center, University of Colorado, Boulder, Colorado

Marvin Newell, Fire Prevention Specialist, Bureau of Land Management, Boise Interagency Fire Center, National Interagency Incident Management System, Boise, Idaho

Capt. Garry Oversby, Inspector Art Contessotto, and Inspector Chuck Gutierrez, Public Information Office, Los Angeles County Fire Department, California

Professors EL Quarantelli and Russel R. Dynes, Co-Directors, Disaster Research Center, University of Delaware, Newark, Delaware

Mike Scherr, Dep. Chief, Fire and Rescue Division, State of California, Governor's Office of Emergency Services; Executive Director, FIRESCOPE, Operations Coordination Center, Riverside, California

For review of the initial drafts:

Capt. WP Carlson, Commander, Sgt. Art Long, and the EMS staff, California Highway Patrol Academy, Bryte, California

Capt. Larry Cloud, Citrus Hights Fire Department, California

Emil Dejan, County Health Services, Sacramento, California

James Holcroft, MD, Associate Professor of Surgery, University of California—Davis, Medical Center, Sacramento, California

Shirley Lowry, RN, Nurse Coordinator, EMT-2 Program, and Paula Madeheim, RN, Nursing Supervisor, Division of Emergency/Critical Care Medicine, University of California—Davis, Medical Center, Sacramento, California

Ed Pelochino, MD, Emergency Physician, Sacramento, California

Harold Renollet, MD, Medical Director, Sierra-Sacramento Emergency Medical Services Agency, California

Special gratitude is owed to Jack and Colleen Files of International Business Systems, Rohnert Park, California, who kept my recurrently malfunctioning computer/word processor working, even loaning me their own when a problem could not be quickly corrected.

CONTENTS

DISASTER RESPONSE
Principles of Preparation and Coordination

CHAPTER 1

THE PROBLEM

The Topeka, Kansas tornado June 8, 1966. (Courtesy of *Topeka Capital-Journal*, Topeka, Kansas.)

Disasters are the ultimate test of emergency response capability. The ability to effectively deal with disasters is becoming more relevant because of factors that tend to increase risk. Unfortunately, there are recurring difficulties with disaster response. Lessons learned in previous disasters are not always being applied in other communities. Sometimes this is because accurate information regarding the basic underlying causes of the difficulties is not readily available to emergency and disaster responders. The purpose of this text is to summarize what the research on disasters has revealed about these underlying causes.

DISASTER: THE ULTIMATE EMERGENCY

"When I got out of the ambulance there were people lying everywhere. Police officers were carrying people out of the front door of the hotel, bystanders were helping others out; and many people were just running into each other trying to get out of the hotel. It was absolute pandemonium. I could see about 200 people outside. Half of them were lying in the grass and the parking lot driveway. . . . When I walked into the hotel, people began pulling at me wanting me to help their wives, husbands or friends. . . . There were people chopped in half, just torsos lying about; people with limbs sheared off, people crushed flat, ones that were still trapped screaming for help. There is no way I can explain the helplessness that overwhelmed me when I saw this. There must have been more than a 100 people still in that hotel dead and in major trauma—and there I stood not knowing what to do next."

Jim Taylor, Paramedic
Hyatt Hotel Disaster
Kansas City
July 17, 1981 (Stout, 1981)

Paramedic Taylor was about to make decisions that could have had life-or-death consequences for a large number of people.

Many people have pondered how they would cope with a disaster of such magnitude. This has been, in fact, a topic of intrigue throughout the ages. Fascination with disasters is reflected in the number of news accounts, movies, and books that deal with the subject. Even before mass communication, and in pre-literate times, ballads sprang up after every catastrophe. People want to know what it is like when a town is obliterated by a wall of water or when high-rise hotel guests are trapped by a fire on the 35th floor.

Figure 1-1.   Rescue operations at the Kansas City Hyatt Regency Hotel skywalk collapse. (Courtesy of the Kansas City Fire Department, Kansas City, Missouri.)

"They want to know how others act under such extreme situations, using disaster as a test for bringing out what is normally hidden—the best and the worst in people, the abilities normal life does not tap, and the dark motives we all repress. As we read about disaster, in safety and comfort, we vicariously test ourselves and our society—How would I have behaved, how would my community have behaved?" (Barton, 1969:xliii)

For the emergency or public safety professional, this question has much more relevance than that of curiosity. These persons are the ones to whom the community will look for leadership and guidance in the event of a disaster. For them, more than anyone else, it is a test of all their training, experience, and ability; for them disaster is the ultimate emergency.

THE RISK OF DISASTER

Concern about disasters is becoming increasingly relevant as increases in population density, population shifts, and increasing technology make it likely that we will encounter disasters more frequently and that they will be more severe (Drabek, 1986:60, 70). Conservative estimates suggest, for example, that the dollar losses from disasters in the year 2000 will be double that experienced in 1980 (Petak, 1985). There are several reasons for this:

Increasing Population Density

For one thing, as areas become more densely populated, there are more potential victims when a disaster strikes. Because of this, future hurricanes or earthquakes of the same magnitude as in the past will tend to result in greater losses (Quarantelli, 1981a).

Increased Settlement in High-Risk Areas

Another reason for increasing disaster losses is that part of this increase in population density is occurring in disaster-prone areas. There is greater settlement in high-risk areas such as flood plains, earthquake faults, coastal hurricane areas, unstable hillsides, areas subject to wildland fires, and areas adjacent to hazardous waste landfills, airports, and nuclear power plants (Petak, 1985; Cigler, 1986; Drabek, 1986:341, 374; Lantis, 1984:2).

For example, between 1960 and 1970 the population along the Texas gulf coast increased 24.8% (as compared to a national population increase of only 14.2%) (Davenport, 1978:1). According to one report, these changes in coastal population density are reflected in the increased time required to evacuate hurricane-threatened areas (17 hrs. for Tampa Bay, Florida; 26 hrs. for Galveston, Texas; and 24 hrs. for Beaumont–Port Arthur–Orange, Texas). This is in the face of estimated maximal warning times of 10 to 12 hours (Ruch, 1984:390).

This pattern of settlement in high-risk areas is reflected in the death rate from flash floods, which tripled from the 1940s to the 1970s (Quarantelli, 1979b). In fact, floods are the most serious natural disaster in the U.S. in terms of lives lost, personal suffering, property damage, and frequency of occurrence. Yet, despite the fact that flood losses are increasing each year, people almost invariably move back into the same area after a flood, and sometimes in even greater numbers (Cigler, 1986; Comm. on Disasters, 1980:20).

> **EXAMPLE:** *Flood, Pearl River, Mississippi, April 11–18, 1979.*
> Property loss in the City of Jackson and its surroundings was
> estimated at $0.5 billion. Yet the most damage occurred to

Figure 1-2.   Burgeoning population along the Gulf Coast without a corresponding increase in transportation routes for evacuation increases the potential for greater loss of life from future hurricanes. Hurricane Betsy struck the southern tip of Florida in 1965 before entering the Gulf of Mexico where it ravaged the Louisiana Coast. (Courtesy of Flip Schulke, Miami, Florida.)

buildings erected in the very same area previously inundated by the flood of 1961 (Drabek, 1985b:32).

Sometimes, in the process of settling high-risk areas, natural protection against environmental threats is removed. For example, along seashores vulnerable to hurricanes, protective sand dunes are removed to make way for houses (Drabek, 1986:374). Paving of large areas in or near settled flood plains prevents water from being absorbed by soil and vegetation. This type of situation in Wilkes-Barre, Pennsylvania, led to extensive flooding during Hurricane Agnes. The property damage that resulted was among the largest in any U.S. disaster (FEMA, 1983d:88).

Finally, the vulnerability of people living in high-risk areas is increasing because the inhabitants are often unaware of the potential risks and how to deal with them (Drabek, 1986:341). People are living in structures that are not designed to resist the forces of local natural hazards. For example, persons moving to the Gulf Coast are increasingly likely to live in mobile homes which are more vulnerable to wind damage, as evidenced by a 700% increase of mobile home sales in these areas in a recent 10-year period (Drabek, 1986:375). Another example is the number of homes built in areas at risk for wildland fires that have extremely flammable, wood-shake roofs.

Figure 1-3.   The recent proliferation of high-rise hotels has increased the risk of disastrous hotel fires. The fire at the MGM Grand Hotel in Las Vegas, Nevada on November 21, 1980 is a good example. (Courtesy of Clark County Fire Department, Las Vegas, Nevada.)

Increased Technological Risks

New technology is adding to the list of disaster agents at an ever-increasing rate. One of these is the 4 billion tons of hazardous chemicals that are shipped annually in this country. Approximately 10% of all trucks (Quarantelli, 1981a) and 35% of all freight trains carry dangerous cargos (Kasperson, 1985). There is a proliferation of high-rise office buildings and hotels that subject their inhabitants to fire threats not experienced before. Our society is also becoming more dependent on technology and specialization, making us more vulnerable to disasters like the mass power outage in the northeastern United States in 1965 (Quarantelli, 1979b; Quarantelli, 1985; Drabek, 1986:375). Our dependence on computers is introducing a new form of disaster vulnerability. A major earthquake in California could disable the computers on which banking and financial institutions depend and which interconnect with the rest of the United States. This could result in a nationwide, or perhaps worldwide, monetary crisis (Quarantelli, 1985).

RECURRING RESPONSE DIFFICULTIES

In the face of ever-increasing risk of disaster losses, there is good reason to be concerned about our ability to deal with these catastrophes. Review of past disasters shows a number of recurring difficulties with disaster response

Figure 1-4.   Unique problems were faced by responders to the Air Florida Crash. U.S. Park Police helicopter rescued survivors clinging to the aircraft wreckage in the Potomac River in Washington, D.C. on January 13, 1982. (United States Park Police Photo—Mr. Charles Pereira.)

though their causes may be considerably more complex than is superficially apparent. One community will experience a disaster and critique its response. Suggestions are made, and an article may be written about the "lessons learned" so that others may benefit from the experience. But, as other communities experience similar catastrophes, the same mistakes are sometimes made all over again. Not only do people sometimes fail to learn from the mistakes of others, but they may even neglect to correct their own, previously noted deficiencies.

> **EXAMPLE:** *The Air Florida Crash, Washington, DC, January 13, 1982.* In a 1980 letter to the Federal Aviation Administration, the Airline Pilots Association stated, "Even though the vast majority of takeoffs and landings at DCA (National) are over water, the marine crash-fire-rescue resources are severely limited." In spite of this warning, the same problems were present when Air Florida Flight 90 crashed into the 14th Street bridge and then into the Potomac river in 1982 (Adams, 1982:54).

ACCESS TO ACCURATE
INFORMATION ABOUT DISASTERS

One of the reasons that lessons about disasters are not learned is because it is difficult for emergency responders and planners to get *accurate* information about what happens in disasters, so that they may profit from the lessons learned by others.

Attributes of Published Disaster Accounts

Many disaster reports lack documentation, objectivity, and perspective. This statement refers, in particular, to those reports in publications most often read by emergency responders.

Documentation

Many articles make recommendations for disaster planning without providing adequate documentation of their validity or effectiveness. Others merely describe a locale's disaster plans or drills, or tell what *"should happen"* rather than what actually *"does happen."* Of the few that describe actual disasters, many of these merely recount what happened rather than analyzing the effectiveness of the plan or the response (Reynolds, 1976:4,9).

Objectivity

Many published articles are narratives of a single disaster written from the perspective of one individual. Frequently, the author is one who was actually involved in the incident or was in charge of some aspect of the disaster planning or response. It is never easy for one to impartially evaluate the actions of his own organizaton. Too often, post-disaster critiques turn out to be defenses or justifications of what was done, rather than objective assessments of problems and mistakes. Valuable lessons may be missed because of the bias held for one's own community or organization (Quarantelli, 1982b:15). In addition, published accounts may delete material that may cause political embarrassment or increase the liability of the response participants. Finally, many disaster critiques are assembled solely for "in-house" use aimed at correcting internal shortcomings and are not meant for others' benefit.

Perspective

The recounting and evaluation of a disaster by a person involved in the response has another inherent limitation, that is, the narrow perspective available to any single participant (especially if his attention is focused on action rather than observation). Each participant can have only a limited view of the total picture (Rosow, 1977:204; Yutzy, 1969:vi). This factor is illustrated by the following quote from an account of the Kansas City Hyatt Hotel skywalk collapse:

> "Nothing we write in this report can describe what happened. The personal accounts provided by EMT Gary Frank and Paramedic Jim Taylor may help the reader gain a better feel for what went on. We were there, and except for occasional flashes of comprehension, we can't get a handle on it ourselves." (Stout, 1981:34)

The task of developing the overall picture of what happens in a disaster is like piecing together the history of a battle. Ernie Pyle, the famous World War II correspondent, observed that:

> "War to the individual is hardly ever bigger than a hundred yards on each side of him." (Pyle, 1946:98)

Of the disaster articles most likely to be read by emergency responders and planners, few are the result of interviews with large numbers of participants in a disaster. Also rare are any articles examining the interrelations among the various responding organizations (Reynolds, 1976:8). This is unfortunate, since coordination among the various agencies involved in the disaster is almost always a problem.

Another limitation of these disaster articles is that they may describe some feature of a disaster, but fail to observe how characteristic it is. Attention may be focused on something because it is particularly dramatic or unusual, even though its frequency of occurrence is small. This may lead to a distorted picture of what is common in disasters and what is not (Barton, 1969:xlvii; Quarantelli, 1982b:15; Quarantelli, 1985:21). The point here is that disciplined study of disasters requires that you make *quantitative* observations. You need to find out such things as: *How often* does panic occur? Or, *how many* casualties have serious (versus trivial) injuries? Or, *how frequently* are there shortages of blood, doctors, or hospital beds? Yet, such quantitative observations are not common in these disaster articles (Barton, 1969:54).

In addition to quantitative observations, *comparative* ones are important. It is essential to determine what disaster characteristics and problems are predictable and recurrent (regardless of the location, type, size, rapidity of onset, and duration of the disaster), rather than just a fluke occurrence in a particular event. However, articles comparing different disasters to see what they had in common—are not common (Reynolds, 1976:2,3,16).

Disaster Research Reports

Because very few emergency responders are likely to be involved in enough disasters to gain adequate personal experience, and because of the narrow perspective offered by personal disaster experience, it is important to be able to collect information on disasters in some organized and disciplined fashion. Only in this way can we hope to gain a more objective idea of what happens in disasters and assemble a reasonable overall perspective of such events. Because this is difficult and expensive, the data have only begun to be accumulated. Nonetheless, a significant body of knowledge has been collected, and it paints a picture about disasters that is sometimes different from what we might expect.

In contrast to many reports in publications read by disaster responders, there are disaster research reports that result from the careful analysis of information collected from a multitude of participants, sometimes even from a number of disasters (Reynolds, 1976; Rosow, 1977:ii; Quarantelli, 1983:15). Unfortunately, despite the fact that a significant body of such information has been collected, much of it is not easily accessible to emergency responders and disaster planners (Quarantelli, 1979b:14,15), because:

1) Much of the research is located in unpublished reports, out-of-print books, and technical journals that are not circulated among emergency responders.

2) Many reports are written using technical terminology that, while appropriate for the academic audience for which they were written, may be difficult for others to understand (e.g., disaster responders and members of emergency response organizations).

For this text, an attempt has been made to survey existing research literature on disasters and to summarize some of the important and relevant concepts and observations. Sources of information include fairly early works, such as the series of studies carried out in the 1950s by the Committee on Disasters of the National Research Council; the University of Chicago, National Opinion Research Center studies (Marks, 1954) and classic works by Fritz (Fritz, 1956; Fritz, 1961), Barton (Barton, 1963; Barton, 1969), Baker (Baker, 1962), Form and Nosow (Form, 1958), Raker (Raker, 1956), and Williams (Williams, 1956).

Much of the more recent research referred to in this text comes from the University of Colorado's Natural Hazards Research Center in Boulder. The largest amount of data, however, are from the Disaster Research Center. The Disaster Research Center was established in 1963 at The Ohio State University, and 21 years later it was relocated at the University of Delaware. It was the first center of its kind in the world and has the largest collection of books, periodicals, and reports related to the socio-behavioral aspects of disasters.

Probably the most interesting and far-reaching study referred to in this text is one carried out by the Disaster Research Center in the 1970s. In this study, E.L. Quarantelli examined emergency medical services in 29 major disasters in various communities throughout the United States and its territories and compared them to identify the common patterns (Quarantelli, 1983). Rather than focusing on the activities of an individual hospital or ambulance service, Quarantelli carried out his observations on a *systems* level, noting how the various organizations affected each other. Quarantelli's works are cited frequently in this text because he has carried out a large part of the existing published research on disasters.

This material is supplemented by a number of technical reports and non-research papers. In most cases, these are used to provide specific examples and anecdotes to illustrate points made in the more academic research studies. Sometimes this was necessary because researchers felt that the subjects they interviewed would be more candid if their names, and the names of their community and organization, remained confidential.

LAG BETWEEN RESEARCH AND PROGRESS

Quality research on disasters takes time, and funding is often difficult to obtain. There is, accordingly, a lag before recent improvements are reflected in the research studies. For example, there are those who may believe that triage has improved since the time of the Disaster Research Center studies (see Chapter 8) of the 1970s. However, another comparative, multi-disaster study of triage has not been carried out to verify that belief. Although there may well have been improvements, some evidence suggests that many of the disaster response problems that were present in the '50s, '60s, and '70s, are still seen in some form in the '80s.

The long list of problems associated with disaster response does not mean to suggest that it is typically carried out by persons or organizations that are incompetent, bungling, or lacking in care and concern. *This is not at all the case.* The history of disasters is rife with unsung heroes, sacrifice, and remarkable improvisation under conditions of *extreme* duress and uncertainty. In many cases, the difficulties are "system problems," not problems with individual behavior or effectiveness. They reflect the fact that organizations evolve to take care of *common* community problems. Disasters, however, pose unique problems often different even from the more routine emergencies that police, fire, medical, and other emergency organizations face on a day-to-day basis. Accordingly, the everyday emergency systems are not always well adapted to tackle disaster problems.

2

THE APATHY FACTOR

An informal party held amidst the rubble of a tornado that struck Topeka, Kansas on June 8, 1966. (Courtesy of the *Topeka Capital-Journal*, Topeka, Kansas.)

Disasters are "low-probability events." As such, they compete for attention with the priorities of daily living. Often, getting the public, elected officials, and organizational leaders to support disaster preparedness is just as difficult as developing the disaster counter-measures themselves. This chapter addresses the causes of apathy toward disaster preparedness, its implications, and some methods for reducing it or planning around it.

WHY IT IS IMPORTANT TO UNDERSTAND APATHY TOWARD DISASTER PREPAREDNESS

One of the social realities to be faced in disaster planning is that the general attitude toward disaster preparedness is characterized by apathy (Drabek, 1987:176). It is important to understand this phenomenon for three reasons:

— to see how it can be influenced,

— to see how it can be circumvented, and

— to develop a realistic appreciation for the limitations it imposes.

REASONS FOR APATHY

Apathy toward disaster preparedness pervades governmental bodies as well as the public at large. Although they will be discussed separately, there is some overlap because governmental priorities are influenced by those of their constituencies.

Public Apathy

Lack of Awareness

Public awareness of disaster risks is generally poor. Even in communities where disasters have occurred relatively frequently, the public has often failed to demand the most rudimentary protection.

> **EXAMPLE:** *Tornado, White County, Arkansas, March 21, 1952.* Even though White County is located in an area known as "Tornado Alley," only about 7% of the people in the impact area had storm cellars, and less than 40% had any knowledge of the appropriate precautionary or protective actions to take in the event of a tornado (Fritz, 1961:661).

> **EXAMPLE:** *Tornado, San Angelo, Texas, May 11, 1953.* Although this city is also located in an area frequently visited by tornadoes, less than 10% of its houses had storm cellars (Fritz, 1961:661).

Figure 2-1.   San Fernando Valley, California, viewed through a breech in the Van Norman Dam after the 1971 earthquake. (Courtesy of the Los Angeles County Fire Department, Los Angeles, California.)

> **EXAMPLE:** *Tornado, Grand Island, Nebraska, June 3, 1980.* No trailer parks had group storm cellars, and the residents had to seek shelter elsewhere (Quarantelli, 1982c:65).

Similar observations can be made in parts of California notorious for earthquakes, where residents fail to anchor tall, heavy items of furniture to the wall and neglect to consider earthquake risk factors when buying property (Drabek, 1986:322; Fritz, 1961:661).

> **EXAMPLE:** *Earthquake, San Fernando Valley, California, February 2, 1971.* An analysis of data collected throughout Los Angeles County after the 1971 quake revealed that few persons had made prior disaster preparations (Bourque, 1973:ii). But, even in the high-impact zone *after* the disaster, less than half the population subsequently made preparations of any type (Drabek, 1986:24).

The motivation for choosing where to live is usually dictated by economic opportunities (a better job) or aesthetic reasons (the risk of wildfire is overridden by the breathtaking view) rather than concern about natural or technological hazards (Drabek, 1986:322,358).

Underestimation of Risk

There is a tendency to underestimate the risk of disaster. It is a striking observation that public *perception* of risk shows no correlation to *actual* risk, and that the risks are usually downplayed.

EXAMPLE: Residents of San Francisco, who are very attached to their city, downplay the risk of earthquake damage. The longer they live there, the less seriously they take the seismic threat. Likewise, 61% of the residents in the flood plain in Tucson, Arizona, do not perceive that they live in a hazardous area. Those who are aware that flooding can occur, underestimate the danger (Drabek, 1985b:4–5; Drabek, 1986: 320).

Sometimes even local legends can contribute to a perception of invincibility:

EXAMPLE: *Tornado, Waco, Texas, May 15, 1953.* An Indian legend held that the area was immune to tornadoes. This was even printed in a pamphlet by the Chamber of Commerce (Moore, 1958:3).

Even direct experience with disasters can inadvertently lead to inaccurate perceptions of risk:

EXAMPLE: Because of their past experience with hurricanes, many persons overestimate their ability to survive future storms by taking minimal emergency measures. Careful scrutiny, however, reveals that many had experienced only the peripheral effects of hurricanes or storms that were not hurricanes at all (Drabek, 1986:324; Davenport, 1978:17).

Reliance on Technology

A false sense of security provided by manmade protective devices contributes to the underestimation of risk. Flood control projects are renowned for this effect. Levees and dams reduce the frequency of flooding and often remove inhibitions against living in a flood plain. Eventually, however, a flood will come that will exceed the capacity of almost any levee or dam. Sometimes, greater settlement in and upriver from the flood plain results in decreased vegetation and more pavement. Eventually, the increase in water runoff will exceed the capacity of the flood control system, leading to catastrophic flooding in a now densely populated area (FEMA, 1983d:88,91; Burton, 1968:13; Drabek, 1986:375).

This problem of a "technological security blanket" is illustrated by the Galveston, Texas, sea wall, constructed in the early 1900s. To most citizens of the city, the sea wall is the "end-all" in hurricane protection. However, even though Hurricane Carla came ashore 70 miles west of Galveston, its tides came within 2 feet of the highest elevation on the island, and the 2 roads to the mainland were under 9 feet of water (Davenport, 1978:3,4,6,17,18).

Fatalism/Denial

The public's attitude toward disasters is often a mixture of "What will happen will happen" and "It can't happen here." People living in high-risk areas accept the threat philosophically ("earthquakes come with the territory") (Drabek, 1986:320, 340).

> **EXAMPLE:** *San Francisco, California, prior to the 1906 earthquake.* "The earthquake back in 1868 had been a ripper. . . . There had been fair shakes in 1892 and 1898, too. . . . 'Nothing to worry about, because there's nothing that can be done about it,' was the attitude. 'Besides, a good shake is not half so bad as a twister or a hurricane bearing down on you' " (Bronson, 1959:19).

Social Pressures

In some high-risk areas, flouting disaster threats is considered a sign of bravery and strong character. This is exemplified by persons going to the beach to party and to surf as a hurricane approaches (Drabek, 1986:340).

> **EXAMPLE:** *Hurricane Carla, Galveston Island, Texas, September 11, 1961.* The attitude of many Galveston Islanders, especially the long-term residents, is one of defiance toward hurricanes. This is typified by the statement of a professional man after Hurricane Carla who said that he was "very proud of not having evacuated. His parents had never fled before a storm . . . , and neither had he." "About 40,000 people (70–80%) stayed on the Island during Carla even though most knew that they would eventually be cut off from the mainland." (Moore, 1964:199; Davenport, 1978:19)

Governmental Apathy

Public apathy, as well as economic restraints, are reflected in a lack of political support for disaster preparedness. Programs have been difficult to initiate or maintain unless they have been demanded by the citizens or mandated by law and paid for by the state or federal government (Tierney, 1985b:73; Davenport, 1978:12). Without federal funding, many government officials have felt that they could only justify the most basic preparedness programs. Sometimes this has been limited to the drawing up of a written disaster plan and assigning the position of disaster coordinator (Seismic Safety Comm, 1979:42; Stevenson,

1981:80). Even when governmental bodies have adopted goals for disaster preparedness, the resources necessary to accomplish the goals have not always been made available (Drabek, 1986:386).

> **EXAMPLE:** *Earthquake and Fire, San Francisco, California, April 18–19, 1906.* In October of 1905, just a bare 7 months before the quake, the National Board of Fire Underwriters had declared San Francisco's 36 million gallons per day water system inadequate. "San Francisco has violated all underwriting traditions and precedents by not burning up; that it has not done so is largely due to the vigilance of the fire department, which cannot be relied upon indefinitely to stave off the inevitable." Fire Chief Dennis Sullivan had battled the supervisors for years to no avail trying to get the money needed to build a supplementary salt-water system and to reactivate dozens of huge, long-neglected cisterns in the city. When the Great Quake finally struck, it was the fire, not the shaking, that dealt the City its greatest blow (Bronson, 1959:21,92).

> **EXAMPLE:** In a study by Wyner and Mann published in 1983, the implementation of earthquake safety measures of thirteen California cities and counties were evaluated. (Seven of these had recent earthquake experiences.) Only a few jurisdictions had attained even their most modest planning goals, which included such elementary accomplishments as collecting more information on the nature of seismic hazards. Most jurisdictions had not even allocated resources in a manner that would permit fulfillment of the adopted goals (Drabek, 1986: 386).

Even when *federal* policy and regulations promote disaster preparedness, these policies are not self-implementing. The enforcement and application of these policies are often dependent on local government cooperation, which is not always forthcoming (Berke, 1987; Clary, 1985:23). At a time when state and local responsibility for disaster countermeasures has increased, they have been confronted by a number of factors that have limited their ability to raise and spend revenues (Mushkatel, 1985:51).

Opposing Special Interest Groups

Vested interest groups can have a negative influence on disaster countermeasures. For example, the most effective way to prevent flood losses is to avoid building structures in the flood plain (Drabek, 1986:353). However, competition for land due to increasing population and the prevailing American attitude that property owners have the right to develop their land without governmental interference, often inhibit restrictions on land use in flood prone areas (Cigler, 1986:10).

Lack of an Organized Constituency Advocating Disaster Preparedness

While special interest groups such as developers, builders, and realtors, are well organized lobbying factions, coalitions promoting disaster prevention and management are not (Cigler, 1986:10, 12, 14; Petak, 1985:5; Clary, 1985:22; Mushkatel, 1985:51, 53; Kasperson, 1985:11; FEMA, 1984a:17).

Defeatism

An attitude of defeatism sometimes contributes to apathy toward disaster preparedness. Some persons believe that every disaster is so unique that effective planning is not possible (Quarantelli, 1979b:2; Quarantelli, 1982b:17; Quarantelli, 1983:104; Quarantelli, 1985:22; Drabek, 1986:58). Others conjure up hypothetical catastrophes of such magnitude as to boggle the mind and paralyze any preparedness effort (Dynes, 1981:iv). Such an attitude may occur when faced with the enormity of preparing for a cataclysmic earthquake or a nuclear holocaust (Seismic Safety Comm, 1979:10; Gratz, 1972:48; Drabek, 1987:55; Blanchard, 1985:3).

Priorities Competing with "Low-Probability" Events

Contributing to governmental apathy is the fact that, in spite of the increasing threat of disasters, they are still improbable events. When crises occur repeatedly at frequent intervals (such as the World War II aerial bombings of London), the target population develops sophisticated mechanisms for dealing with the threat (e.g., air raid wardens and bomb shelters). Peacetime disasters occur so rarely, however, that there is not usually the impetus to make elaborate, time-consuming, and expensive preparations (Barton, 1969:40; Dynes, 1981:72). Those areas that have the most extensive disaster preparedness are the ones exposed to recurrent seasonal threats from floods, hurricanes, and tornadoes (Drabek, 1986:55,178; Barton, 1969:40).

This improbability of occurrence is especially true with respect to *large-scale* disasters in the United States. In fact, only six disasters in U.S. history have resulted in more than 1,000 fatalities (see Table 2-1) (Quarantelli, 1987; Lane, 1984; Houghton, 1986; Grolier, 1985), and only about 10 or 15 disasters per year have resulted in more than 40 injuries (Wright, 1977:193).

As stated by one renowned disaster researcher, the statistical probability is

Table 2-1.   U.S. Disasters with Deaths Exceeding 1,000

1. April 27, 1865. Steamship Sultana explosion on the Mississippi River near Memphis, Tennessee. 1,547 killed.
2. October 8, 1871. Forest fire, Peshtigo, Wisconsin. 1,182 deaths.
3. May 31, 1889. Flood, Johnstown, Pennsylvania. More than 2,200 deaths.
4. August 27, 1900. Hurricane, Galveston, Texas. About 5,000 deaths.
5. June 15, 1904. Fire on the steamship General Slocum, on the East River, New York. 1,021 fatalities.
6. September 13, 1928. Hurricane, Lake Okeechobee, Florida. 2,000 deaths.

that when a disaster strikes, it will strike elsewhere—primarily because there is so much "elsewhere" (Drabek, 1985b:4). From 1900 to 1967 there were only 16,619 deaths from natural disasters in the United States (Roth, 1970:442). These figures pale when compared to more routine causes of violent death. For example, in 1967 alone, the number of persons dying in motor vehicle accidents was 52,924 (Nat Safety Council, 1973:12). The infrequency of disaster is reflected in the figures shown in Table 2-2.

Because of the improbability of disaster impact, the expense and effort put out to prepare for it is perceived as an investment with little certainty of return (Barton, 1969:159). In the face of the doubtful benefits of preparing for a catastrophe that may never occur, are competing everyday concerns such as employment, crime, pornography, attaining status among one's peers, or even partaking of leisure activities (Fritz, 1961:661; Drabek, 1985b:4; Drabek, 1986:320). In other words, people are unlikely to give priority of attention to an unlikely future disaster when there are fifteen tasks that have to be accomplished by Friday (Tierney, 1985a:77). This factor is particularly salient in contemporary government where there are so many programs competing for scarce resources (Drabek, 1985b:i; Seismic Safety Comm, 1979:42; Blanchard, 1985:4).

> **EXAMPLE:** When state and local decision-makers were asked to rate the importance of 18 problems that might require governmental attention, the highest ratings were given to inflation, welfare, unemployment, and crime. The lowest ratings were given to floods, hurricanes, tornadoes, and earthquakes (Drabek, 1986:385).

The mention of these factors should not be taken to imply that, because disasters are improbable, effective countermeasures are not practical. Rather, it is to point out that motivational issues need to be considered during disaster planning.

Figure 2-2.   (**A**) Galveston, Texas, prior to the hurricane of 1900 and after (**B**) the hurricane. No domestic peacetime disaster has caused more fatalities than this storm. (Courtesy of the Rosenberg Public Library, Galveston, Texas.)

Table 2-2.

Risk	Fatalities per person-hour of exposure
Natural disaster	1 in 100 billion
Smoking	5 in 10 billion
Motor vehicle transport	1 in 1 million

[Adapted from Foster and Starr (Foster, 1980:19; Starr, 1969)]

Difficulty Substantiating Benefits of Preparedness

Assessing risks from potential disaster hazards is difficult at best, as is determining the benefits of disaster management and preparedness efforts. This is complicated by the expenditures required to make the necessary studies and the uncertainty, even then, that the answers are forthcoming. Ironically, the very apathy that inhibits disaster preparedness is just as likely to thwart the funding of studies to assess the cost-benefit ratio of disaster preparedness (Petak, 1985:5,6; FEMA, 1984a:47,125; Cigler, 1986; Kasperson, 1985:9,10; Zimmerman, 1985:33). Often, the benefits of preparedness are not visible in the short run, but only after a disaster has occurred. In times of economic restraints, programs whose benefits cannot clearly be demonstrated get short shrift on the list of budgetary priorities.

Overestimation of Capability

Another reason for complacency toward disaster preparedness is the mistaken belief that the disaster problems can be managed merely by an extension of routine emergency measures. In studying disaster emergency medical services, the Disaster Research Center found this attitude in over half the communities in their sample (Seismic Safety Comm, 1979:9; Quarantelli, 1981a:10; Quarantelli, 1983:101; Sorensen, 1981:27; Barton, 1969:159). However, as discussed in Chapter 4, disasters often pose unique problems for which routine emergency procedures are not well adapted.

The Inter-governmental Paradox

As you move to lower levels of government, the disaster damages experienced from that level's perspective are less frequent. For example, the federal government experiences most all of the disasters that occur in the nation. State government experiences fewer of them, and city governments even fewer. Because any given local government experiences the fewest exposures to disaster loss, it is least apt to perceive it as an important issue (Cigler, 1986:8,13; Drabek, 1985b:4).

The inter-governmental paradox refers to the fact that the local government, which is least likely to see disaster management as a key priority, is most likely to be faced with the responsibility for carrying out the disaster response. There are several reasons for this. For one thing, the trend in the United States has been to place most of the responsibility for disaster response on local government (Cigler, 1986:8,10; Clary, 1985:23,24). For another, local governments bear a large part of the responsibility because they are the closest to the event and are apt to be on the scene before substantial state or federal resources are available (Kasperson, 1985:13).

Ambiguity of Responsibility

Responsibility for disaster tasks in the United States is spread out among many public and private organizations (Drabek, 1987:105). In addition, disasters tend to cross jurisdictional and functional boundaries involving city, county, state, federal, and special district (e.g., flood control or fire districts) governments as well as private spheres of responsibility. This often results in a situation where no *single* institution, person, or level of government is perceived as responsible for disaster preparedness. Accordingly, disaster preparedness goals and policies of various jurisdictions and agencies are often contradictory, and motivation to get things done is hampered by a lack of accountability (Cigler, 1986:6).

TAKING APATHY INTO ACCOUNT WHEN PLANNING

Apathy results in limited resources for disaster preparedness. It is important to apply the limited interest and resources in such a manner that they will do the most good. In effect, *there is a form of "triage," or priority-setting, that has to be applied to disaster preparedness measures,* and the cost-effectiveness of these measures has to be taken into consideration.

PRINCIPLE

Because of the limited resources available, disaster preparedness proposals need to take cost-effectiveness into consideration.

Prepare for What Is *Likely*

Moderate-Sized Disasters

Some disaster planners believe the best planning philosophy is to prepare for the worst-case disaster scenario. They anticipate disasters involving tens of thousands of casualties (Drabek, 1985b:i; Dynes, 1981:75). A variation on this theme is disaster planning aimed primarily at nuclear attack (Gratz, 1972:48). At present, preparedness for nuclear attack, rather than the more common types of natural and technological disasters, is the top planning priority of the Federal Emergency Management Agency (Thomas, 1988:14). There are, however, several problems associated with planning for disasters of this magnitude:

- They may conjure up hypothetical possibilities of such immensity as to make most despair at ever being able to cope with them (Dynes, 1981:iv,75; Blanchard, 1985:3). In other words, such scenarios amplify apathy.

- Most of the research on disasters in the United States has focused on relatively moderate events involving tens to hundreds of casualties. The applicability of this knowledge to cataclysmic disasters involving tens and hundreds of thousands of casualties is questionable. Although such "cataclysmic" disasters have occurred in foreign countries, the socio-political context is so different from the United States that it is unclear whether we can apply lessons learned from them to our own culture and form of government. In other words, while we now know a fair amount about how to plan for smaller disasters, we know little about how to plan for the cataclysmic ones (Drabek, 1985b:3; Drabek, 1986:6). *Therefore, when it comes to applying limited resources to planning, it makes sense to plan for those events for which we have the knowledge to plan.*

- Applying limited resources to prepare for the most improbable type of disaster—the cataclysmic one—is not cost effective. It is also not the best approach when we have not first become proficient at handling small disasters, which are reasonably likely to occur. It is like signing up for very expensive dance lessons before learning to walk.

For these reasons, preparedness for moderately sized disasters may be more realistic and achieve greater acceptance by those who must pay for and carry out the preparations (Lewis, 1980:865; Dynes, 1981:75). This is *not to say that planning for large disasters isn't valuable.* Clearly, California *will* experience a large earthquake in the near future. However, planning is sometimes carried out for cataclysmic disasters *to the exclusion* of the more moderate and more likely ones.

The advantage of a focus on moderate disasters is that the procedures involved are more likely to be *used* and, therefore, learned. They are also more likely to get funded. Furthermore, the skills, training, procedures, and supplies developed for moderate disasters are a logical step toward preparedness for larger events.

On the other hand, a number of the preparedness efforts for cataclysmic disasters have never been used and have little applicability in the types of disasters more likely to occur. An example is the packaged disaster hospitals. These have laid dormant for so long that you may have to question if the equipment and supplies are still usable, or if those who would need them know that they exist, where they exist, how to get them, what they contain, or how to set them up.

Just what is a moderately sized disaster? That answer is less clear. Limited evidence suggests that a disaster large enough to consistently cause inter-organizational coordination problems is one resulting in about 120 casualties (Wright, 1977:190). Since inter-organizational coordination is one of the major disaster response problems, this figure seems to be a reasonable yardstick.

> **PRINCIPLE**
>
> Planning should be for disasters of moderate size (about 120 casualties); disasters of this size will present the typical inter-organizational coordination problems also applicable to larger events.

More Common Types of Disasters

Disaster preparedness applicable to the more common disaster events is more likely to receive support. It is easier to sell planning for multiple vehicle accidents than for an earthquake. Serious earthquakes occur in the same locality once in several decades; complex traffic accidents occur several times a year. Likewise, tornadoes, floods, and hurricanes are a more relevant threat than a nuclear holocaust. Apathy is most likely to be overcome by emphasizing those threats that are seen as relevant by the public and elected officials (Drabek, 1987:203; Stevenson, 1981:40).

Predictable Disaster Tasks

Although it is not possible to prepare for every disaster contingency, there are some problems that occur with such regularity as to be quite predictable. It is these which are the most amenable to planning. For example, almost every major disaster requires procedures for the centralized gathering and sharing of information about the overall disaster situation and the responding and available resources. Procedures are needed for overall coordination (deciding what organizations are going to carry out what tasks and how they are going to interact) and for logistics such as supplies, transportation, feeding, shelter, and communication networks to support the disaster response. Other examples include procedures for integrating and managing unsolicited volunteers, warning threatened populations, handling evacuations, carrying out triage, coordinating search and rescue, keeping unauthorized persons out of the impact area, distributing casualties rationally among the available hospitals, decontaminating equipment and casualties exposed to hazardous materials, dealing with the press, and responding to voluminous inquiries from anxious loved ones and relatives of those thought to be potential disaster victims. This idea of focusing on "generic" disaster tasks most likely to be faced in all types of disasters has been embodied in the concept of "comprehensive emergency management" which the Federal Emergency Management Agency (FEMA) has used in its "Integrated Emergency Management System" (FEMA, 1984c:I-9; FEMA, 1983a; FEMA, 1985d:1–2).

Some of the tasks likely to be needed can be identified if a formal hazards analysis is carried out. Available methods for doing this are well described in several publications available from FEMA (FEMA, 1983b:9; FEMA, 1984c:II-10; FEMA, 1983a:3–2; FEMA, 1983c:11; FEMA, 1985c; FEMA, 1985b; FEMA, 1985a). However, it must be realized that present methods of risk analysis are crude at best and can be used as only a rough guide for disaster planning (FEMA, 1984a:47,125). (Appendix A illustrates a method of hazards analysis.)

Make Preparedness Adaptable to a Wide Variety of Circumstances

Adapt to Routine Emergencies

Whenever possible, it is advantageous to adapt disaster procedures for use in daily, routine emergencies (Div Med Sci, 1966; Morris, 1982; Jenkins, 1975). This has several benefits: 1) it may reduce training costs by decreasing the number of procedures that are used only in disasters; 2) it allows those who will need to use the procedures to become familiar with them; and 3) it may improve routine emergency responses (Stevenson, 1981:4,42,44; Quarantelli, 1983:149).

One example is the emergency medical services system in Sacramento County, California. The standard practice there is for emergency medical technicians on the ambulances to assign a triage category to every emergency patient transported on a daily basis. This becomes a part of the radio report to the hospital. The result is that both the emergency medical technician and hospital personnel become familiar with the triage system. In essence, they have a daily triage drill (Lowry, 1983).

Modular Expansion

The flexibility of a disaster plan is enhanced if the disaster management structure is designed to be expanded in stages as the incident (and the number of resources that need to be coordinated) grows in size. If managed well, this can help to minimize the presence of excessive numbers of personnel and organizations and, therefore, simplify coordination (Tierney, 1980:100). (See Chapter 7.)

Cost-Sharing

Joint training, combined dispatch centers, standardized resource and management procedures, and joint purchasing of standardized equipment can all be promoted as cost-reduction measures applicable to routine emergency responses. They also have been shown to facilitate multi-agency coordination and communication in disaster operations, discussed further in Chapter 5.

REDUCING APATHY

Although apathy is difficult to overcome, it is possible to motivate disaster preparedness.

> **EXAMPLE:** As the result of a research project on earthquake-related fires, it was projected that San Francisco would be swept by firestorms in the event of another 1906-like earthquake. With the encouragement of researcher Charles Scawthorn, and buttressed by their own estimations, the San Francisco Fire Department declared that there was a pressing need to increase the city's preparedness. With the support of the mayor, the fire department conducted an information campaign to explain the fire risk and the necessary countermeasures to the public. The media were involved, and a citizens' committee was formed to initiate a bond issue. In November of 1986, they were successful in obtaining a 90% approval of the voters for a $46.2 million bond issue to finance improvements in the city water system, firefighting capability, and an emergency operations center (NHRAIC, 1987a:13).

There are a number of motivating factors that can be used to counteract apathy. Some illustrative examples follow.

Liability

Changing liability related to disasters may help to counter apathy. Recent court decisions indicate that local governments and officials may be financially liable for certain consequences of a disaster if the community was not prepared or did not respond properly (Tierney, 1985b:58; Stevenson, 1981:83; Drabek, 1986:342; Perkins, 1984). With such a change, emergency managers can rightly argue for increased resources to meet these increased legal responsibilities (Adams, 1981b:51).

The trend suggests that governments will increasingly face the threat of successful litigation if someone suffers damage from a disaster. A loss due to disaster can be recovered in court if the victim can show: 1) the governmental body owed a duty to the victim to avoid, prevent, or lessen such a loss; 2) that the body failed to carry out this duty; and 3) that the loss occurred as a result of this failure (Kusler, 1985:119).

There are several factors that may increase governmental liability in this context (Kusler, 1985:120):

- The courts have recognized broadened concepts of the duties and responsibilities of local governments.

- The "act of God" defense for disaster losses is less frequently accepted by the courts.

- The ability of governments to claim "sovereign immunity" ("The 'king' can do no wrong.") has been substantially reduced.

- The duty of governments to develop disaster countermeasures is becoming more frequently stipulated in legislation.

Recent Disasters

One factor, more than any other, appears effective in reducing apathy—the occurrence of a disaster. This theme surfaces repeatedly in the research literature on disasters. Unfortunately, the interest so generated also decays rapidly, often before it has had time to stimulate significant changes (Drabek, 1985b:5; Drabek, 1986:366; Quarantelli, 1983:138; Seismic Safety Comm, 1979:45; Stevenson, 1981:81).

> **PRINCIPLE**
> Interest in disaster preparedness is proportional to the recency and magnitude of the last disaster.

If there is a preparedness program lacking support, one should be ready to take advantage of a disaster to reintroduce it—even if the disaster has occurred elsewhere (Stevenson, 1981:82). While broad-based appeals for support can be based on such tactics, they are most effective if tied to specific needs, with this illustrative disaster exemplifying that need (Drabek, 1987:176).

> **EXAMPLE:** One preparedness director in South Dakota used a movie of a disaster that struck another community. He showed his county commissioners the film, "The Day of the Killer Tornadoes," in an attempt to obtain funding for an emergency generator. The film graphically depicted a blackout in the stricken community's Emergency Operations Center. As a result, the commissioners voted unanimously to budget for the generator (Drabek, 1987:175).

> **EXAMPLE:** *Earthquake, Whittier, California, October 1, 1987.* "Perhaps one of the reasons we have accomplished so much in earthquake mitigation in California is that whenever concern over seismic threat ebbs for too long, the earth rumbles and shakes and warns us that we had better take this threat seriously. As much as we in the mitigation business try to be proactive, to foresee problems, threats, and dangers, and to mitigate their effects, we are not entirely successful when we are working in an atmosphere of near apathy. We try to combat the apathy in the 'off years' by producing some excitement, holding pretend earthquakes in the back lot of Universal Studios, with the tremendous visual and sound effects only the studio can re-create (sic). These events attract media and citizen attention, and they exercise and practice our own response operations. Needless to say, however, their impact pales in comparison to a real trembler. . . . "

> "Immediately following the earthquake, the usual apathy and inertia were overcome, if only for a brief time. Both the public and government officials were receptive to mitigation messages. Everyone wanted to be briefed, educated, and trained. . . . "

> "The day after the quake, councilman Hal Bernson introduced a motion to create within the fire department a division to supervise training the community to prepare for earthquakes. . . . Remarkably, this very costly program, originally proposed more than 2 years before, was unanimously approved a few weeks after the quake. . . . " In addition several

other council motions were taken, and council action sparked
renewed interest in improving the safety and security of the
Emergency Operations Center (Mattingly, 1988).

> **PRINCIPLE**
> The best time to submit disaster preparedness programs for funding
> is right after a disaster (even if it has occurred elsewhere).

Public Education

Public education about the nature of disaster hazards and the practical countermeasures available can help to offset apathy. Education is most effective at
times when people are motivated to learn about disasters. For example, the
public is more anxious to learn about disasters at the beginning of seasonal
threats (e.g., tornado or hurricane season), or after disasters, even non-local
ones, that have received attention in the news (e.g., the Mexico City earthquake). At these times, they may be more interested than usual in how vulnerable their own community is to disaster threats, how well their emergency
services are able to respond, and what practical measures citizens can take to
protect themselves. A listing of sources for public education and awareness
material is given in Appendix E.

The Media

One of the most effective methods of public education is the mass communications media, particularly commercial television and radio stations (Holton,
1985:17; Wenger, 1985b:17). News accounts emphasizing the lack of and need
for improved emergency preparedness increases support for larger operating
budgets and helps to make preparedness a higher local government priority
(Stevenson, 1981:36). Tornado, blizzard, hurricane, and earthquake media
awareness releases have been used effectively to improve public awareness of
disasters (Drabek, 1987:198; FEMA: A-70).

School Programs

School programs teaching about disaster hazards increase awareness, not only
in the students, but in their parents as well. Evaluation of a junior high school
program on hurricane awareness in North Carolina found that approximately
80% of the students obtained parental help in the program's homework assignments (FEMA: A-26). An article in the *Journal of Civil Defense* (12(2):10–13, 1979)

gives an illustration of how adults can be influenced indirectly by disaster education directed at children:

> **EXAMPLE:** In 1974, as a school bus was transporting children home, a tornado approached. Although the driver did not know what to do, a seventh-grade student on the bus did. His teacher had reviewed tornado precautions in class that day. The student convinced the driver to stop the bus and get everyone into a ditch. Although the tornado destroyed the bus, none of the pupils was injured (Foster, 1980:187).

A National Weather Service study found that the subject of disaster preparedness is most appropriate for the fifth, sixth, and junior high school grades. It also recommended that the curriculum should not be costly and should contain material addressing local types of hazards as well as those of a more general nature. The material for school programs was more likely to be accepted by schools if it was introduced within existing school curriculum such as social studies or science courses (FEMA:A-74).

SUMMARY

Disasters are "low-probability" events. As such, they are associated with a high degree of apathy. It is important to understand the limitations posed by this fact. The existence of apathy should not be taken as an excuse to neglect or discount the need for preparedness, but, especially in this time of shrinking resources and expanding responsibilities, we must be selective in deciding which aspects of disaster preparedness to emphasize. Programs that are practical, inexpensive, realistic, and applicable, not only to a wide variety of disasters, but also to routine emergency problems, are the most likely to gain support. Even then, however, motivation for disaster planning is one of the most difficult aspects of the planning and preparedness process.

PLANNING CHECKPOINTS

[ ] Does your disaster plan focus on events of moderate size? Ones which are likely to occur in your area?

[ ] Does your disaster training program emphasize *common* disaster tasks, such as overall coordination, determining what the overall disaster problems are, determining all of the resources present and available, decontamination, search and rescue, patient distribution, management of volunteers, handling inquiries about disaster victims, and relations with the press?

[   ] Have you adapted disaster procedures for application in routine emergencies, so personnel can become familiar with them?

[   ] Can your disaster response plan be expanded in stages as the incident magnitude increases?

[   ] Does your community disaster planning include provisions for cost-sharing of resources and training?

[   ] Are your elected officials and organizational leaders aware of the potential liability of failing to develop disaster countermeasures?

[   ] Do you have public education material prepared for dissemination:
— at the beginning of seasonal disaster threats?
— after the occurrence of major disasters?

[   ] Do you have arrangements with the media for broadcast of disaster education material?

CHAPTER

3

THE "PAPER" PLAN SYNDROME

Disaster plans are an illusion of preparation unless accompanied by training. (Courtesy of Vern Paule, Public Information Officer, FEMA Region IX, San Francisco, California.)

Written disaster plans are important, but they are not enough by themselves to assure preparedness. In fact, they can be an illusion of preparedness if they are not tied to training programs, not acceptable to the intended users, not tied to the necessary resources, or not based on valid assumptions. This illusion is called the "paper" plan syndrome. This chapter discusses the important steps in avoiding impotent written disaster plans.

WHAT IS THE "PAPER" PLAN SYNDROME?

One of the greatest impediments to disaster preparedness is the tendency to believe that it can be accomplished merely by the completion of a *written* plan (Quarantelli, 1982b:16; Quarantelli, 1985:21). Written plans indeed are very important, but they are *only one* of the requirements necessary for preparedness (Gratz, 1972:12; Quarantelli, 1981a:12; Bush, 1981:1). A written plan can be an illusion of preparedness if the other requirements are neglected (Quarantelli, 1982b:16,17; Rosow,1977:104; Barton,1963:43; Barton,1969:96; Moore, 1958:10). This illusion will be referred to as the "paper" plan syndrome.

"PAPER" PLANS VERSUS DISASTER RESPONSE

The preponderance of "paper" plans is reflected in the frequency with which disaster responses differ from what is in the written plan (Neff, 1977:181; Golec, 1977:175; Worth, 1977:160, 162; Rosow, 1977:104, 105; Quarantelli, 1983:87,121; Moore, 1958:21; Tierney, 1985b:62; Arnett, 1983:31; Dynes, 1981:71).

> **EXAMPLE:** Many hospital administrators concede that while disaster plans are necessary for hospital accreditation, they are relatively unworkable in practice (Worth, 1977:166). As stated by one administrator involved in a disaster: "I opened up our plan immediately after we were notified, and it said that wards 4A and B would be the shock and resuscitation areas for all victims. That's four floors up. I've got two old elevators that take forever to move up, and I said we're forgetting the disaster plan completely, this is the way we are going to run it, and we ran it from that point on our own. . . . " (Worth, 1977:166)

> **EXAMPLE:** *Mt. St. Helens Volcano Eruption, May 18, 1980.* A Washington State University study revealed that a majority of 26 communities did not use an emergency preparedness plan when the eruption occurred. In many cases, city officials discovered that the plan was not applicable to their needs (FEMA, 1983d:8).

A Disaster Research Center study of 29 mass casualty disasters found that in most cases the disaster plan was not followed to any great extent. One reason for this was that key personnel did not fully understand the plan or know their role in it. In addition, common disaster problems were not anticipated. In only 21% of the disasters was a pre-designated communications plan followed, and in less than 50% of the cases was transportation of disaster casualties carried out according to the written plan (Quarantelli, 1983:71,89,121).

PRINCIPLE

Disaster planning is an illusion unless: it is based on valid assumptions about human behavior, incorporates an inter-organizational perspective, is tied to resources, and is known and accepted by the participants.

PLANNING BASED ON VALID ASSUMPTIONS ABOUT HUMAN BEHAVIOR

Disaster planning and response must be based on valid assumptions. Unfortunately, many of the assumptions people make about disasters are incorrect (Quarantelli, 1982b:15; Quarantelli, 1985:3,19,21; Drabek, 1985b:i,9).

The Myth of Maladaptive Behavior

One of these assumptions is that citizens in a disaster-impacted area tend to respond in a maladaptive manner. One common belief is that panic is a common occurrence and that warnings and evacuation orders must be given most cautiously to prevent it. Another belief is that many persons are stunned by the impact and suffer from a condition of immobility and inability to act rationally (the so-called "disaster syndrome"). These persons are thought to be incapable of acting on their own and to need strong leadership and direction by authorities. Another belief is that the chaos and confusion following disasters provides the conditions for antisocial behavior such as crime, looting, and exploitation (Dynes, 1974:71; Quarantelli, 1960:68; Quarantelli, 1965:107; Quarantelli, 1972:67).

The prevalence of belief that disasters are typified by maladaptive behavior is suggested by the results of a study by Wenger and his associates (Wenger, 1975). They surveyed a random sample consisting of 354 residents of New Castle County, Delaware. Of those surveyed, 84% believed that panic is a major problem in disasters, 74% felt that disaster victims cannot care for themselves because they suffer from the "disaster syndrome," 62% felt that looting was usually a significant problem in disasters, and 51% believed that crime rates usually increase immediately following disasters.

A later study by Wenger, James, and Faupel confirmed these results. This study included the previous data from New Castle County (a community with little actual disaster exposure) and added random sample surveys of the general population from three communities that had suffered from multiple major disasters (300 interviews from each community). A total of 51 additional interviews were carried out with informants from emergency response organizations in these communities (including the mayor's office, civil defense, police, fire, sheriff, Red Cross, Salvation Army, military, and hospital). This study revealed that most emergency responders also held these beliefs, though not as large a percentage as the general population (Wenger, 1985a:103,105).

Although seemingly less prevalent in more recent publications, the belief in maladaptive behavior is still expressed in articles and books on disaster management (FEMA, 1983a:5–16; Buerk, 1982:644; 1981f:40; Arnett, 1981:76,87).

Careful and systematic studies of disasters, however, have yielded an entirely different picture. Although an occasional episode of human behavior may conform to this stereotype, it does not represent the ways in which people typically respond to disasters (Mileti, 1975:57; Quarantelli, 1960:68; Quarantelli, 1965:107; Wenger, 1975:33; Quarantelli, 1972:67). The morals, loyalties, respect for laws, customs, and tenets of acceptable behavior, ingrained by years of upbringing, are not dissolved in an instant by disaster (Drabek, 1968:143). Courage, altruism, and selflessness are characteristics far more representative of disaster behavior (Drabek, 1986:143). (The prevalence of helping behavior in disasters is discussed in Chapter 6.) As stated by Professor E.L. Quarantelli of the University of Delaware's Disaster Research Center:

> "Most human beings act in quite controlled and adaptive ways
> in the face of the new and extreme stresses which they face
> during large scale disasters." (Quarantelli, 1965:108)

Panic is not a typical response to disaster. On the contrary, it is often quite difficult to get persons in a disaster-threatened area to evacuate. (More discussion about panic is found in Chapter 9.) Furthermore, disasters generally do not render people stunned and unable to act. They will take what they perceive to be appropriate actions even without direction or leadership from the authorities. In fact, official directives that are not considered relevant or appropriate may be ignored altogether (Quarantelli, 1960:76; Dynes, 1974:30; Fritz, 1956:41; Fritz, 1961:672).

Except in civil disorders, it has been difficult to verify that significant looting or an increase in criminal activity occurs in peacetime disasters. The investigations that have been carried out conclude that looting is quite rare, and that criminal activity does not increase (Quarantelli, 1972:69; Drabek, 1986:145, 180;

Dynes, 1968:10; Fritz, 1957:53). In a study of 100 disasters, researchers found many stories of looting, but extremely few verified cases (Dynes, 1981:26; Quarantelli, 1972:69).

> **EXAMPLE:** *PSA Air Crash, San Diego, California, September 25, 1978.* After the airliner collided with a private plane and crashed into a residential area, a report of looting at the crash site was circulated. The San Diego police chief was so concerned by this unverifiable rumor, that he wrote a letter to a national news magazine, stating, "There is absolutely no evidence that any looting occurred at the crash site or in the immediate vicinity." (Drabek, 1986:146)

> **EXAMPLE:** *Tornado, White County, Arkansas, March 21, 1952.* Of those questioned by a University of Chicago team of investigators, 58% stated that they had heard of *others'* property being stolen, but only 6% felt convinced that *their own* property had been taken. Furthermore, most of the missing items were of inconsequential value. The study team could actually verify the theft of only two items—a cash register and a piano (Quarantelli, 1972:69; Dynes, 1968:10).

Unfortunately, police sometimes invest so much effort in preventing looting that traffic control suffers. Serious crowd and traffic problems interfere with access, and movement of ambulances and rescue equipment are not managed optimally (Kennedy, 1970:358). However, because the public *expects* looting and other anti-social behavior, they need to feel the presence of security forces in the area. It is not usually necessary in natural disasters to deploy *massive* forces for this purpose. Rather the need can often be met by placing a few armed guards at strategic and conspicuous locations, and by mass media announcements that all necessary precautions are being taken (Dynes, 1981:33). Furthermore, the deployment of security and law enforcement personnel for traffic and perimeter control also contributes to their visible presence to the public.

"Likely" Behavior Versus "Correct" Behavior

Disaster plans are often written in the belief that people ought to behave according to the plan. The plans state what people *"should do."* A more successful approach is to design the plan according to what people are *"likely to do."* Plans are much easier to change than human behavior (Drabek, 1985b:9; Quarantelli, 1985:21; Dynes, 1981:iv).

> ## PRINCIPLE
> Base disaster plans on what people are "likely" to do, rather than what they "should" do.

Some planners believe that persons in the impact area are often stunned and rendered helpless by a disaster and that what they need and will respond to is a strong leader who can tell them what to do. For this reason, it is assumed that disaster behavior can be controlled to a high degree. Some disaster plans specify in quite some detail the manner in which people are to behave or respond.

However, even totalitarian governments using coercive measures during wartime have not always found it possible to dictate behavior that was not considered legitimate by the public. When this has been tried, orders have sometimes had to be rescinded in the face of widespread and sometimes violent protests.

> **EXAMPLE:** *Germany, World War II.* The government had to abandon attempts to prevent families from bringing their children back to the cities which were targets of allied bombing attacks (Quarantelli, 1960:76).

> **EXAMPLE:** *Britain, World Wars I & II.* The British government in both world wars tried to ban the use of subway stations as overnight bomb shelters. Both times, however, people continued to sleep in the subway stations, and regulations against this activity had to be canceled (Quarantelli, 1960:76).

Some persons seem at times resistant to evidence that contradicts their presumption that control of disaster behavior can be achieved. The failure of citizens to follow their directives may be interpreted to reflect a weakness in the means of control used, rather than in the basic assumption that such control can be achieved. Occasionally, disaster officials gain a false sense of success when they misinterpret public actions as being a result of their directives (Quarantelli, 1960:77).

> **EXAMPLE:** Warning was received in a California city that it might be the target of a tidal wave. City officials issued an order to evacuate the downtown area. The evacuation order was called a success because the area was cleared rather promptly. However, many of those who left then went to the beach to watch for the wave! (Quarantelli, 1960:76)

INTER-ORGANIZATIONAL PERSPECTIVE

EXAMPLE: "The key to NASA's success in reaching the moon was that all the participants were impressed not only with their role in getting the rocket off the ground but more importantly with how their role interfaced or interacted with other roles. They were briefed not only on their duties but also informed about the total, overall project. The problem of getting to the moon was solved by many experts performing in their own separate fields of expertise but all with the same goal in mind. Although each participant had only a small role in the outcome, each was very much aware of his own part in achieving it." (Coleman, 1978:8)

This quote was taken, not from a book on space exploration, but from a fire management text. The author used it to describe the importance of an overall *systems* perspective in fire service operations. A systems perspective in disaster preparedness requires *inter-organizational* planning. Some of the most critical difficulties in disaster response are due to the lack of inter-organizational coordination. Yet, many organizations plan for disaster as if they were to function in isolation. Their disaster plans are conceived with a focus on trees rather than forests.

For example, while nearly all hospitals have disaster plans, they may have ignored coordinating them with other hospitals, public safety agencies, and ambulance services (Quarantelli, 1983:103; Worth, 1977:166). The Disaster Research Center found that only 44% of the communities they studied had any inter-organizational disaster plan whatsoever for emergency medical services. Even then, some plans called for the coordination of only two or three emergency agencies. Plans called for police to coordinate with fire departments, or for ambulances to coordinate with hospitals, but other organizations were ignored.

Furthermore, most of the plans only took into consideration those emergency organizations that normally respond to medical emergencies within the political boundaries of the community. Even fewer plans existed for overall coordination of disaster emergency medical services at the county or state level (Quarantelli, 1983:86,120; Neff, 1977:179; Tierney, 1985a). In only about 25% of the localities did the Disaster Research Center find any type of regional disaster planning (Quarantelli, 1983:106).

EXAMPLE: *The Air Florida Crash, Potomac River and Subway Derailment, Washington, D.C., January 13, 1982.* The National

Transportation Safety Board investigation revealed that there was no area-wide disaster plan that provided for joint response by emergency units of the District of Columbia and adjoining suburban areas of Virginia and Maryland. The D.C. Fire Department and the Transit Authority had jointly conducted three disaster drills prior to the derailment. One of the drills was an evacuation of 292 passengers from the subway. However, the simulations did not include participation by suburban fire and rescue units, D.C. Police, or the metropolitan area hospitals (NTSB, 1982:46).

EXAMPLE: *Earthquake, Coalinga, California, May 2, 1983.* All agencies involved in the response to Coalinga did have disaster plans. However, most of the plans were not coordinated with those of other agencies and jurisdictions, and when the quake struck, the various organizations seemed to act independently. Poor coordination among the responders resulted in misunderstandings, delays, and duplication of effort (Seismic Safety Comm, 1983:74; Kallsen, 1983:29; Tierney, 1985b:33).

PRINCIPLE
For disaster planning to be effective, it must be inter-organizational.

There are two types of organizations in particular that are frequently overlooked in community disaster responses. The first of these is the military. The second is the private sector, especially private hospitals (Stallings, 1971:28,30; Hildebrand, 1980:12).

REALISTIC SUPPORT FOR DISASTER PLANNING
Resources Necessary to Carry Out the Plan

One of the reasons that disaster plans may not be put into effect when disaster strikes is because of the failure to provide the resources (personnel, time, money, equipment, supplies, or facilities) necessary to make the plan work (Barton, 1969:96).

EXAMPLE: *Hyatt Hotel Skywalk Collapse, Kansas City, Missouri, July 17, 1981.* Although the use of a medical emergency triage tag (METTAG) was designated by the Kansas City disaster plan, the necessary materials were not available the night of the disaster. Therefore, triage tags were not used. Similarly, identification arm bands prescribed in the disaster plan were not available (Orr, 1983:602,603).

Plans may be developed, but funding not made available for equipment and supplies. Time and money may not be budgeted for the development of disaster training programs. Many emergency organizations operate on a 24-hour-a-day basis. This means that ongoing training sessions must be repeated for each shift, or personnel must come in on their day off (but overtime pay for this may not be budgeted). Persons may be assigned disaster planning tasks, but not given the paid time to carry them out effectively and still meet their routine work obligations. Rather, they may have to donate their free time. Few rewards and little recognition are provided to induce knowledgeable and experienced persons to become involved in disaster preparedness activities. It is little wonder that communities that allow planning to occur in this context—get "what they pay for." If disaster planning is to result in more than "paper" plans, the planning process must be tied to the resources necessary to carry out the mandate (Drabek, 1986:386; Seismic Safety Comm, 1979:42; Kilijanek, 1981:41; Dynes, 1981:74; May, 1985:45; Mushkatel, 1985:51).

Status of the Disaster Planning Office

In order to gain the attention, respect, and cooperation of other governmental offices, disaster planning must be given a place in the governmental hierarchy that provides the necessary status, authority, and support (Stevenson, 1981:42). Unfortunately, disaster planning is often relegated to a position of low status in the administrative hierarchy of organizations—isolated from any existing sources of political power and from the priority-setting, budgeting, and decision-making processes (Drabek, 1986:53; Tierney, 1985b:74).

There is a theoretical advantage when the disaster preparedness office functions as a staff position to the governmental chief executive officer, independent of other governmental subdivisions. When community disaster planning is relegated to a single agency such as the fire or sheriff's department, its priorities sometimes take second place to those of the agency. Also, cooperation with other agencies can be dampened because the disaster office is not seen as a neutral body.

However, such an "independent" position is not always the most advantageous. Sometimes an individual sheriff or fire chief can offer support, legitimacy, and authority to disaster planning which more than offsets the theoretical advantages of an "independent" disaster planning office. A wise preparedness director will seek a niche for his agency that provides the strongest base of support. The exact location of this niche will vary from one community to another (Tierney, 1985b:74; Drabek, 1987:194,233).

INVOLVEMENT OF DISASTER PLAN USERS

Knowledge of the Plan

Disaster preparedness cannot be accomplished unless the plan is known by the participants (Quarantelli, 1981a:17; Adams, 1981a:25). History has shown us the consequences of this fact.

> **EXAMPLE:** *The Great San Francisco Earthquake and Fire, April 18–19, 1906.* The third floor of the fire station on Bush Street was the official residence of Chief Dennis Sullivan. When the quake struck, it toppled a set of brick smokestacks which plummeted through the roof, critically injuring the Chief. He was taken unconscious to the Southern Pacific Hospital and died 3 days later. San Francisco will never know what might have happened if the Chief had not been injured. He, more than any man in the city, had been aware of the frightful fire potential presented by the miles of crowded wooden buildings. Apparently unbeknownst to anyone else, he had long before laid plans to stop the kind of conflagration that could result if the city's water supply were disrupted. There was water in the bay, and there were ways to pump it into the city (Bronson, 1959:29, 40).

> **EXAMPLE:** *The Evacuation of Mississauga, Ontario, Canada, November 10, 1979.* After the derailment of a Canadian Pacific Railway train in 1979, chlorine leaking from one of the tank cars made it necessary to evacuate 220,000 residents of Mississauga, including three hospitals and several nursing homes. The evacuation was successful because of the pre-planning, training, and experience of the Peel Regional Police Force. The Peel Police plan was not only a good plan, but it was known to the members of the force. This was because of the requirement that all officers know the plan in order to pass promotional examinations (Drabek, 1986:120; Quarantelli, 1982a:H-36).

Training Programs

Disaster plans, in order to be functional, must be tied to training programs (Casper, 1983; Quarantelli, 1985a:21; Dynes, 1981:75). It is during training sessions and drills that various operational problems can be encountered and resolved (Adams, 1981a:25). Unfortunately, although 83% of local governments have disaster plans, only 52% actually test these plans (Mushkatel, 1985:51), and only 42.2% of counties and 27.7% of cities test them annually (Drabek, 1985a:86).

Agencies are often more easily motivated to participate in practical simulations and training programs than to expend valuable resources developing rigid and complex written plans whose value they question. Practical courses such as those put on by the Federal Emergency Management Agency's Emergency Training Center or the California Specialized Training Institute are but a couple of examples (Seismic Safety Comm, 1979:26).

Designation of Positions Rather Than Persons

It is important that participants in the disaster response know how to carry out the plan even in the absence of certain key individuals. Therefore, plans should be written in terms of positions (for example, the on-call administrative supervisor, or the acting chief), rather than in terms of particular persons. Succession of authority should be covered by the plan (for example, who is in charge if the mayor is out of town) (Worth, 1977:166; Quarantelli, 1983:121).

> **EXAMPLE:** *Earthquake, Coalinga, California, May 2, 1983.* Bob Semple was Coalinga's public information director and a volunteer emergency medical technician. When he got back to his office, he had to dig through the rubble for his copy of the county disaster plan. The first thing it said to do was to find the incident commander, who was supposed to be the Coalinga Fire Chief. Unfortunately, the Fire Chief was out of town. Using his car CB radio, Semple did manage to contact the captain in charge of the fire department, and they began to try and get things organized. (It was a Fresno County Sheriff's lieutenant who subsequently assumed the position of incident commander.) (Arnett, 1983:31)

Acceptance of the Plan

According to a 1979 report, city managers and county executives feel that state and federal disaster agencies require the writing of very complicated and lengthy disaster plans. City managers said they had read the plan once, did not know where it was now, and wouldn't use it in a disaster anyway (Seismic Safety Comm, 1979:19). As stated by one city official:

"Once you get your plan approved by the Office of Emergency Services, go bury it and write a plan which meets your needs." (Seismic Safety Comm, 1979:19)

And an emergency services coordinator put it this way:

She pointed to a 3-inch volume on the shelf and said that it was the official city disaster plan. She called it a "compliance plan"—a term that was used by emergency services directors around the state to describe the plan that had been submitted to the Office of Emergency Services. From the desk drawer she withdrew a thin handbook, stating that this was the city's real plan. This consisted of a list of agencies, contacts in the agencies, telephone numbers, a list of where to get various kinds of equipment and supplies, and a checklist of actions to take in various types of disasters (Seismic Safety Comm, 1979:19).

The point to be made is that disaster plans must be acceptable to the elected officials, the departments that will implement them, and even to those the plan is intended to benefit [the public]. The consequence of ignoring this principle is that the resulting plans may also be ignored (Gratz, 1972:39,50).

Importance of the Planning *Process*

One aspect of disaster planning often overlooked is the importance of the *process* (Drabek, 1986:53; Wenger, 1986:72). Often it is more important than the written document that results. One reason for this is that those who participate in developing the plan are more likely to accept it. This is preferred over adopting a plan written by someone else who may not understand local circumstances. But, there is another aspect of equal importance—the personal contacts that develop. A number of researchers have observed that pre-disaster contacts among representatives of emergency organizations result in smoother operations in subsequent disasters. Organizations are more likely to interface if the contact is not with total strangers. Furthermore, in the process of planning, the participants become familiar with the roles of other individuals and organizations involved in the disaster response (Dynes, 1978; Drabek, 1986:125; Quarantelli, 1983:120,130). (See Chapter 5)

> **PRINCIPLE**
> The *process* of planning is more important than the written document that results.

Planning *by the Users*

One of the reasons disaster plans become "paper" plans is because they are often composed by civil defense officials or disaster planning offices rather than by the emergency agencies that have to carry out the response (Gratz, 1972:48). This pattern has its roots in the historical wartime focus of civil defense in the United States (Blanchard, 1985). Planning for wartime civil defense was based on the assumption that local emergency response agencies would not be acting as independent and autonomous bodies, but would act as part of a "military-like," line organizational structure under the direction of the federal government. With an emphasis on an enemy attack scenario, civil defense planning was based on military experience, and civil defense officials were often appointed who had a military background (Irwin, 1984).

Even after the "dual use" concept (preparedness applicable both to enemy attack and peacetime disasters) was introduced in the early 1970s, civil defense planning was seen as an effort to get local governments to comply with federal policy. In order to receive federal civil defense money, local government had to comply with complex paperwork requirements and create written disaster plans according to specific rules. Since the federal perspective was based on a military orientation, little of the required planning dealt with issues critical to the realities of emergency response in the civilian context.

It is within this historical context that planning is still often seen as something that is done *by* a civil defense office (or what is now often called an emergency management agency) *for* the community's emergency response organizations (e.g., fire, law enforcement, ambulances, hospitals, Red Cross). This type of planning effort is sometimes enhanced when a disaster advisory committee composed of representatives of local emergency response organizations is formed. Unfortunately, it is still often seen as a plan imposed from the outside. For this reason, its legitimacy and effectiveness may be questioned by those for whom it is intended (Wenger, 1986:13,60; Drabek, 1987:60,62,106,178; Dynes, 1978:52; Gratz, 1972:48).

Recently, a different organizational structure for disaster planning has gained in popularity. This model is represented by a congressionally funded project called FIRESCOPE (Firefighting Resources of Southern California Organized for Potential Emergencies) which was made up of federal, state, and local firefighting agencies in Southern California. FIRESCOPE was chartered in 1972 after a series of devastating wildland fires. Its purpose was to develop coordination processes for multi-agency fire operations. The important feature of the FIRESCOPE process is planning *by the users* (the responders). This process has been adopted by the National Interagency Incident Management System (NIIMS) for use by federal wildland firefighting agencies on a national basis (ICS, 1983b; ICS, 1986; FEMA, 1987:5).

Planning group membership in this model is open to representatives of all those organizations likely to be participating in local emergency operations.

The planning process actually describes a four-tiered decision-making and planning body (see Table 3-1):

— **The Board of Directors** is made up of agency directors and sets goals and policy.

— **The Operations Team** is composed of the agency operations chiefs (deputy, assistant, or division chiefs, those who are directly in charge of each agency's emergency operations). They implement Board decisions and recommend new proposals for consideration or review.

— **The Task Force** is composed of supervisory operations-level officers (for example, fire battalion chiefs, police sergeants). They provide most of the general staff work and basic analysis.

— **Specialist Groups** are composed of agency specialists (for example, experts in training, communications, public information). They perform technical staff work in their areas of expertise.

Administrative management of the planning and implementation process is through a **Coordinator** who is responsible to the Board of Directors. The coordinator selected should be as free as possible from the influence of any single agency or jurisdiction. If he is unduly influenced by a particular agency,

Table 3-1.   The FIRESCOPE structure for emergency and disaster planning

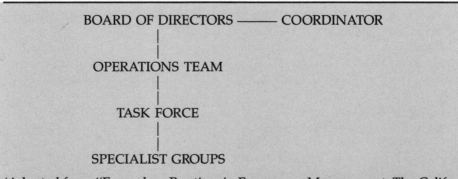

BOARD OF DIRECTORS ——— COORDINATOR
|
|
OPERATIONS TEAM
|
|
TASK FORCE
|
|
SPECIALIST GROUPS

(Adapted from "Exemplary Practices in Emergency Management: The California FIRESCOPE Program," Monograph series No. 1, Federal Emergency Management Agency, Washington, D.C., 1987, p. 5.)

his credibility and effectiveness will be compromised (ICS, 1983b; Wenger, 1986:68).

The FIRESCOPE planning process is designed so that the jurisdictional authority and responsibilities of the participating agencies are not compromised. The planning group attempts to clarify each member agency's roles and how they will interact with other member agencies (Irwin, 1988). Among the other advantages of this approach, this "planning by the users" assures that the resulting plan will be known and accepted by those who are supposed to put it into action.

Because FIRESCOPE was mandated by Congress to address the response problems of *fire agencies*, this model was designed primarily for fire service planning. Although it was not specifically designed for the participation of representatives of multiple disciplines such as fire, law enforcement, hospitals, and military organizations, nor of elected chief executives such as mayors, city managers, county supervisors (Irwin, 1988), the model is easily adapted to include representation by these participants. One format for elected officials might be a joint powers body (FEMA, 1983d:159) made up of a chief elected executive or his representative, from each participating political jurisdiction (e.g., city mayors, county supervisors, special district supervisors, state, and federal representatives). The board of directors would then answer to that joint powers body. This body would set overall political policy and establish intergovernmental agreements regarding budgetary support.

SUMMARY

Disaster plans are not effective unless several requirements are first met. They must be based on valid assumptions about what happens in disasters and how people tend to behave when faced with such crises. Disaster plans must also take a "systems" perspective. They must take into account all of the organizations and persons involved in the response, even the unexpected ones. Finally, disaster plans must be familiar to those that will use them, and accepted by them as legitimate and appropriate. Plans that do not fit these criteria may only succeed in creating a false sense of security in the community for which they are written. In contrast to the traditional approach to disaster planning, where the civil defense authorities establish planning requirements for the responders, there is a new and more effective model. This new approach is for the directors of the agencies themselves to determine their needs and to establish multi-agency coordination arrangements. The coordinator for this planning effort is selected by the agency directors and acts on their behalf. This approach tends to assure that planning corresponds to local needs and that the resulting plan is accepted and understood by those who will need to use it.

PLANNING CHECKPOINTS

[ ] Does your disaster plan cover the tasks and responsibilities of *all* the organizations and individuals likely to be involved in the response?

[ ] Does it avoid trying to change how people normally behave in disasters?

[ ] Does your planning body or coordinator have the respect, support, and authority to carry out his mandates? Is he chosen by the agencies that will be using the plan?

[ ] Are those who are expected to implement and use the plan familiar with it?

[ ] Do the involved organizations have ongoing, mandatory disaster training programs?

[ ] Do those for whom the plan is designed accept it as legitimate and worthwhile? Did they *develop* the plan?

[ ] Are disaster tasks assigned in terms of positions rather than individuals?

[ ] Does your disaster planning office have a position within the governmental hierarchy where its input is likely to be heard by those who set operational and budgetary priorities?

ADDITIONAL READING

Public official attitudes toward disaster preparedness in California, Publication No. SSC 79-05, 1979. Available from: Seismic Safety Commission, 1900 K St, Suite 100, Sacramento, Calif 95814 (free).

4

DISASTERS ARE DIFFERENT

Disasters can damage emergency response resources. This photograph shows the collapsed ambulance bays at Olive View Hospital, a result of the San Fernando Valley earthquake of February 9, 1971. (Courtesy of the Los Angeles County Fire Department, Los Angeles, California.)

One of the reasons disaster response is difficult to coordinate is because *disasters are different from routine, daily emergencies*. The difference is more than just one of magnitude. Disasters generally cannot be adequately managed merely by mobilizing more personnel and material. Disasters may cross jurisdictional boundaries, create the need to undertake unfamiliar tasks, change the structure of responding organizations, result in the creation of new organizations, trigger the mobilization of participants that do not ordinarily respond to local emergency incidents, and disable the routine equipment and facilities for emergency response. As a consequence of these changes, the normal procedures for coordinating community emergency response may not be adapted well to the situation.

WHAT *IS* A DISASTER?

What is a disaster? The term often suggests images such as earthquakes, tidal waves, floods, hurricanes, and explosions, and yet it is difficult to define a disaster by physical characteristics alone. Is the flooding of an uninhabited, uncultivated plain a disaster? What about a landslide in a deserted canyon? In general, to be considered a disaster, these events have to affect an area of human development.

Often, even this is not enough. An earthquake might cause little damage in California, because the target area has relatively earthquake-resistant buildings. The same amount of seismic activity in a foreign community, whose buildings have unreinforced stone walls, might result in disastrous loss of life. Thus, definition of a disaster must include consideration of a hazardous event's effect on the target population.

The impact of a hazardous event on a community is partially determined by the mechanisms and adaptations that the population has developed to deal with the effects of potentially damaging events. In some communities, natural hazards occur with such regularity that effective methods have been developed to cope with them. In such cases, the event might trigger emergency activity, but would not result in a disaster.

Imagine the impact of a typical Vermont winter on Southern Florida, or a monsoon season in Phoenix, Arizona. Vermont does not declare a disaster every winter, because the residents have adapted to the weather there. Cincinnati, Ohio, is an example of a city which has adapted to frequent flooding. Located in the flood plain of the Ohio River Valley, Cincinnati is subject to flooding about every 14 months. As a result, local organizations have developed a sophisticated set of procedures for responding to floods. These are so effective that flooding emergencies do not inordinately disrupt the community's coping mechanisms (Anderson, 1965).

The amount of property destruction and numbers of deaths and injuries are often used as a criteria for defining a disaster, but this may be somewhat misleading. A ten-victim, multiple vehicle collision might overwhelm a rural community hospital, whereas the same event may not cause undue problems at a large urban trauma center. Research does suggest that non-routine procedures and inter-organizational coordination are almost always required when a civilian disaster produces more than 120 casualties (Wright, 1977:190). In contrast, much more death, injury, and loss of material are managed in wartime without exceeding the ability of the system to respond effectively and smoothly. Bomb shelters, fire control, management of debris clearance, and systems for handling the dead and wounded become routine (Yutzy, 1969:36).

Some disaster plans identify three levels of disaster. A typical version is described in *The Student Manual for Disaster Management and Planning for Emergency Physician's Course* (ACEP:1-2):

Level I: A localized multiple casualty emergency wherein local medical resources are available and adequate to provide for field medical treatment and stabilization, including triage. The patients will be transported to the appropriate local medical facility for further diagnosis and treatment.

Level II: A multiple casualty emergency where the large number of casualties and/or lack of local medical care facilities are such as to require multi-jurisdiction (regional) medical mutual aid.

Level III: A mass casualty emergency wherein local and regional medical resource's capabilities are exceeded and/or overwhelmed. Deficiencies in medical supplies and personnel are such as to require assistance from state or federal agencies.

These definitions of disaster levels can be useful for planning different levels of disaster response, but one caution is in order. It must be recognized that *even in local (Level I) disasters*, federal and state agencies are often involved, and need to be considered when coordination procedures are planned.

The Federal Emergency Management Agency defines disaster as:

"An occurrence of a severity and magnitude that normally results in deaths, injuries, and property damage and that *cannot be managed through the routine procedures* and resources of government. [Emphasis is author's.] It usually develops suddenly and unexpectedly and requires immediate, coordinated, and effective response by multiple government and private sector organizations to meet human needs and speed recovery." (FEMA, 1984c:I-3)

Holloway, a physician who has written a number of articles on disaster management, defines a disaster as:

"Many people trying to do quickly what they do not ordinarily do, in an environment with which they are not familiar." (Tierney, 1985a:77)

This requirement, to do things in non-routine ways, often under conditions of extreme urgency, is one of the keys to understanding disaster response problems. Often, to a significant degree, disaster-stricken communities end up *improvising* their responses.

A common disaster planning assumption is, "Good disaster response is merely an extension of good, routine, daily emergency procedures." (Quarantelli, 1981a:10; Quarantelli, 1983:87; Tierney, 1977:153; Orr, 1983:603; Gratz, 1972:48; ACEP:9-2; Sorensen, 1981:27) In contrast to this assumption, this chapter will discuss the ways in which disasters differ from more routine, daily emergencies, and why *the normal ways in which communities cope with routine emergencies may not always work well in disasters.* These differences are not limited to questions of magnitude. While it is true that the destruction posed by a disaster is often greater than that from routine emergencies, there are also differences in the types of problems that must be handled, the types of tasks that must be carried out, and the types of help available. Thus, responding to a disaster involves more than merely mobilizing greater numbers of emergency personnel and greater quantities of supplies.

> ## PRINCIPLE
> Good disaster management is *not* merely an extension of good everyday emergency procedures. It is more than just the mobilization of additional personnel, facilities, and supplies. Disasters often pose unique problems rarely faced in daily emergencies.

ROUTINE PATTERNS OF EMERGENCY MANAGEMENT

The management of routine, daily emergencies in the United States is influenced by a national preference for local control and private enterprise. The result is a complex, decentralized structure where the various tasks are divided up among a myriad of independent public and private organizations and individuals (Drabek, 1985a:85; Drabek, 1987:105; Quarantelli, 1981c:68). Which organization does what at the scene of an emergency is usually determined by tradition and is formalized in laws, contracts, and charters. The geographical areas to which each emergency agency responds and the roles and responsibilities of each are often mutually understood. Occasional jurisdictional disputes do occur, but these are usually settled by legislative or judicial procedures, or by informal negotiations over a period of time.

Since local emergencies usually involve the same set of emergency organizations, each is eventually able to carry out its tasks at the scene independently and with *relatively* little conflict or confusion. Because these tasks often do not tend to change, there is frequently not a great need for on-the-spot decisions about the responsibilities of each organization at the scene. In short, routine emergencies create little demand for ongoing, moment-to-moment coordination *among* the involved organizations (Quarantelli, 1985:5; Dynes, 1978:59; Dynes, 1981:12,39; Wenger, 1978:27).

As in the assignment of tasks, the assignment of resources for routine emergencies is often standardized. Each organization is budgeted a quantity of resources, and this may be done far in advance of their use—often on an annual basis. Members of each organization may be familiar with the needed resources, where they are located, and the standardized procedures for obtaining them. In such cases, much of the information regarding the availability of resources is known in advance of an emergency response and does not need to be communicated.

IN DISASTERS, THE DIVISION OF LABOR AND RESOURCES CHANGES

In disasters there are often conditions that may make the traditional division of labor and resources, characteristic of routine emergency management, unsuitable for disaster response:

- Disasters may put demands on organizations, requiring them to make internal changes in structure and delegation of responsibilities.
- Disasters may create demands that exceed the capacities of single organizations, requiring them to share tasks and resources with other organizations that use unfamiliar procedures.
- Disasters may attract the participation of organizations and individual volunteers who usually do not respond to emergencies.
- Disasters may cross jurisdictional boundaries, resulting in multiple organizations being faced with overlapping responsibilities.
- Disasters may create new tasks for which no organization has traditional responsibility.
- Disasters may render unusable the normal tools and facilities used in emergency response.
- Disasters may result in the spontaneous formation of new organizations that did not exist before.

Table 4-1.   Differences in Disasters

Routine Emergencies	Disasters
Interaction with familiar faces	Interaction with unfamiliar faces
Familiar tasks and procedures	Unfamiliar tasks and procedures
Intra-organizational coordination needed	Intra- and inter-organizational coordination needed
Roads, telephones, and facilities intact	Roads may be blocked or jammed, telephones jammed or non-functional, facilities may be damaged
Communications frequencies adequate for radio traffic	Radio frequencies often overloaded
Communications primarily intra-organizational	Need for inter-organizational information sharing
Use of familiar terminology in communicating	Communication with persons who use different terminology
Need to deal mainly with local press	Hordes of national and international reporters (see Chapter 10)
Management structure adequate to coordinate the number of resources involved	Resources often exceed management capacity

Organizations Change Internally

Emergency response organizations may adapt to the increased demands of a disaster in a number of ways which can result in their members carrying out unfamiliar tasks with unfamiliar equipment and interacting with unfamiliar faces (Dynes, 1974:81; Dynes, 1978:50; Quarantelli, 1978:4; FEMA, 1983d:14).

Off-Duty Personnel Called In

Emergency organizations such as hospitals, ambulance companies, fire departments, and law enforcement agencies typically operate 24 hours a day. Also, some emergency organizations such as police departments may have a cadre of reserve or auxiliary officers that can be summoned for duty (Quarantelli,

1972:69). By calling in off-duty personnel, the available manpower may be quickly doubled or tripled. Unfortunately, this may also deplete the reserve of well-rested personnel if the disaster lasts longer than one work shift (Quarantelli, 1983).

Personnel Re-assigned to New Duties

Organizations may curtail non-essential activities and re-assign personnel to disaster-relevant duties (Dynes, 1981:44,62). Fire departments may re-assign fire prevention officers to fire suppression duties. Police departments may re-assign detectives, training officers, crime prevention officers, and records personnel. In addition, routine patrol activities may be reduced. Hospitals may discontinue routine services like: elective surgery, clinic services, patient education, physical therapy, medical library services, and gift shop hours (Dynes, 1981:44,62).

Everyday Procedures and Priorities Altered

In disasters, emergency organizations are often required to use different procedures and to establish different priorities for action. One example is the hospital, where medical treatment may be carried out in different areas of the facility and by different personnel than usual. Nurses sometimes end up making medical decisions, such as which patients to discharge to make room for disaster victims. Suspected fractures may be splinted without being X-rayed. Arriving patients may have been exposed to dangerous chemical or radioactive material and require decontamination. Physicians, nurses, medical students, and student nurses who do not usually work in emergency treatment areas may be pressed into service there. Registration of incoming patients may be abandoned in favor of using disaster tags. In many cases, record-keeping and billing are abandoned in favor of more rapid treatment and disposition. Hospital switchboard operators are often inundated with unusual offers of assistance or requests for information. The phone lines may quickly become so jammed that it becomes impossible to use them to get information into or out of the facility. This introduces the novel task of using alternative means to carry out communications. A system of runners may have to be set up to carry messages. Hospital security is faced with new tasks related to traffic and crowd control (Quarantelli, 1983:82,83; Worth, 1977:164; Tierney, 1985b:33,80; Williams, 1956:658; Stallings, 1971:18).

Persons manning communication and decision-making positions may become so overwhelmed with the volume of traffic that they are forced to perform a sort of "**communications triage.**" That is, they must filter out all but the most essential information to transmit. A problem can occur when the person filtering the information does not understand its significance to the overall disaster effort (Stallings, 1971:18; Kilijanek, 1979:5; Dynes, 1977:10,12; Brunacini, 1985:47). This may be because the information is important to another organization whose goals and tasks are unfamiliar (Ringhofer). Confusion can also occur when other persons (those, for example, from a subdivision of the organization that has ceased to have a priority function during the disaster) are pressed into manning radios or answering phones.

Overload of communications channels and filtering of information in disasters can have widespread effects on decision-making. Getting information to higher echelon authorities in organizational bureaucracies can become too time consuming and unreliable for the situation at hand. Decisions have to be made urgently or lives and property are lost. The result is that the decision-making in disasters may tend to occur at lower levels in the organization than in routine emergencies (Drabek, 1986:121,162,171; Dynes, 1977:6; Dynes, 1978:60; Drabek, 1985a; Drabek, 1985b:20; Tierney, 1985b:32, Stallings, 1971:32; Worth, 1977:163; Rosow, 1977:74).

Organizations Share Tasks and Resources

Even with the various adaptations available to them, *single organizations* often do not have the resources to manage the disaster tasks at hand. It may not be possible for all the traffic to be controlled by one police department, all of the rescue and extrication to be carried out by one fire department, all the injured to be transported by one ambulance company, and all the patients to receive timely medical care at one hospital. One of the fundamental differences in disasters is that various urgent tasks may have to be divided up among multiple organizations. In contrast to the more common, large emergencies in which mutual aid is requested from familiar neighboring jurisdictions, organizations sharing tasks in disasters may be from distant locales and may have not worked together before. Pre-existing mutual aid agreements or familiar procedures for working together and sharing resources may be lacking (Drabek, 1981:21; Quarantelli, 1985:5; Dynes, 1974:79; FEMA, 1983d:14; Quarantelli, 1983:64; Kilijanek, 1981:126; Quarantelli, 1981a:10; Dynes, 1981:41).

Involvement of Non-emergency Responders

Many organizations and individuals that normally do not do so on a routine basis may become involved in emergency tasks. Some organizations have traditional mandates to switch to emergency-related activities in the event of a disaster. Examples are the Red Cross, the Salvation Army, public works departments, television stations, and private utility companies. Non-emergency governmental organizations such as parks departments, purchasing departments, and building inspectors may also be pressed into disaster activities. There are others with no such mandate, but they become involved because of the perceived need and a spirit of altruism. In addition to all the individual volunteers that become involved, there are many organizations that donate their services. Examples include labor unions, church groups, scouts, civic and fraternal organizations, private helicopter operators, and heavy construction companies (Dynes, 1974:136).

Crossing of Jurisdictional Boundaries

Disasters often involve the response of many independent organizations from the private sector as well as from multiple levels of government, including federal, state, county, and city agencies, as well as special districts (such as fire districts, regional parks, and water districts). The diversity of responders is illustrated in Table 4-2 adapted from a study of search and rescue operations in disasters (Drabek, 1981).

Government in the United States is very decentralized. According to the 1982 Census of Governments, there are over 82,000 separate governments in this country. Such decentralization allows for, and in some cases even promotes, a lack of standardization. This is illustrated by the variations in the way authority is vested to activate local disaster plans, to request state disaster assistance, and to order a major evacuation. Even the organizational structures of local disaster agencies are characterized by diversity (Drabek, 1985a:85; Drabek, 1987: 107, 233; Wenger, 1986:59). This lack of standardization among the myriad of agencies representing various levels of government complicates coordination of disaster response.

Unfortunately, many organizations continue to act independently in disasters, focusing on their own organizational tasks, and sometimes failing to see or find out how their role fits into the *overall* response effort. This has been called by some the **"Robinson Crusoe syndrome"** ("We're the only ones on the island."). This narrow focus on one's own organizational goals has been observed not only in disaster response, but in planning as well. In a number of communities, the various organizations that have a role in disaster response have carried out their planning individually with little attempt to meld their plans together into a coherent overall strategy. This problem has been noted even more frequently with planning on a county- or state-wide basis. Different levels of government (city, county, state, federal, special district) may not have plans which are coordinated with each other (Quarantelli, 1983:87,103,120; Tierney, 1985a:73; Tierney, 1985b:33; NTSB, 1982:46; De Atley, 1982:33; Adams, 1982:54; Kallsen, 1983:29; Kilijanek, 1981:44; Neff, 1977:179; Seismic Safety Comm, 1979:56; Seismic Safety Comm, 1983:71).

Table 4-2.   Organizations Involved in Search and Rescue

Disaster	Private	City	County	State	Federal
Tornado Lake Pomona, KS 6/17/78	5	4	5	4	2
Flood Texas Hill Country 8/1–4/78	3	2	13	6	1
Tornado Wichita Falls, TX 4/10/79	4	5	5	4	2
Tornado Cheyenne, WY 7/16/79	4	3	4	3	1
Hurricane Frederic Jackson County, MS 9/12/79	3	8	6	3	1
Volcano Eruption Mt. St. Helens, WA 5/18/80	2	1	7	5	10

(Adapted from: Drabek T.E., Tamminga H.L., Kilijanek, et al., "Managing Multiorganizational Emergency Responses: Emergent Search and Rescue Networks in Natural Disaster and Remote Area Settings," Natural Hazards Information Center, University of Colorado, Boulder, 1981.)

OBSERVATION

The typical response to a disaster includes multiple independent organizations from the private sector as well as from agencies of city, county, state, federal, and special district governments. Often, they have planned independently and end up responding that way, with little grasp of how each fits into the *overall* response.

When planning has been done on an inter-organizational basis, it is more likely to result in a coordinated response.

EXAMPLE: *Tornado, Wichita Falls, Texas, April 10, 1979.* The city, county, and state had well rehearsed and detailed disaster plans. They were designed to fit with one another and to be complementary. The general roles and authority structure were understood by most. Following impact, the response network formed very rapidly according to the previously practiced plans (Adams, 1981b:30,40).

Political Boundaries

Earthquakes, wildfires, tornadoes, hurricanes, floods, and toxic spills may cause destruction over large geographical areas, simultaneously involving city, county, regional, state, and federal jurisdictions. Under such circumstances, customary divisions of responsibility may be inapplicable (Quarantelli, 1985:16; Quarantelli, 1983:106; Neff, 1977:179; Tierney, 1977:154).

EXAMPLE: A railroad tank car containing a toxic volatile substance exploded at a chemical plant within the limits of a northeastern city. However, the gas cloud that resulted spread into the county area (Quarantelli, 1983:58).

EXAMPLE: *Volcano eruption, Mt. St. Helens, Washington, May 18, 1980.* This disaster involved a large federal jurisdiction (U.S. Forest Service) and that of three counties. Further jurisdictional overlap resulted when the Governor of Washington declared the event a state disaster (Drabek, 1981).

Examples of the governmental agencies that might be involved at various levels are shown in Table 4-3.

Disasters do not need to cover large geographical areas in order to cross multiple levels of governmental responsibility. Even localized disasters can include federal jurisdiction if a navigable waterway or airport is involved. Jurisdictional boundaries may be crossed even when geographical boundaries are not. For example, the federal government may have jurisdictional involvement in a local accident involving a nuclear reactor or the crash of a military aircraft. Laws determining who has overall coordinating responsibility and authority when jurisdictions are crossed are often unfamiliar to the participants, or are vague, confusing, or non-existent (Sorensen, 1981:46).

EXAMPLE: *Tornado, Lake Pomona, Kansas, June 17, 1978.* This tornado struck and capsized a showboat on a federal reservoir in a state park in an unincorporated area of the county. Unfortunately, there was no existing state law which defined who should be in charge of such a situation. (Even if one existed, it

Table 4-3.   Governmental Agencies Involved in Disaster Response

Law Enforcement and Investigatory Agencies	
City	*Federal*
Police	National Guard
	Secret Service
	Bureau of Alcohol, Tobacco, and Firearms
County	
Sheriff	National forest special agents
Coroner	Park Police
Park ranger	Fish and Wildlife Service
	Coast Guard
State	Bureau of Indian Affairs
Police, highway patrol	Environmental Protection Agency
Fish and game wardens	Nuclear Regulatory Commission
State forest/park ranger	Department of Transportation
University police	Aviation Administration
	Highway Traffic Safety Administration
	Railroad Administration
	Maritime Administration
	FBI
	Other
	Fire department arson investigation bureaus

Fire Protection Agencies	
Local	*Federal*
City fire departments	Forest Service
Local fire protection districts	Department of the Interior
County fire departments	National Park Service
	Bureau of Indian Affairs
	Bureau of Land Management
State	
State forestry	

Medical Organizations	
V.A. hospitals	Military air-sea rescue
County hospitals	State and county health offices
Public Health Service hospitals	State emergency medical services offices
Military hospitals	

cont'd.

Table 4-3. Governmental Agencies Involved in Disaster Response *cont'd.*

Medical Organizations	
Public ambulance and rescue teams	U.S. Park Service mountain rescue
Lifeguards	County sheriff's search and rescue
Military land ambulances	teams
	Civil Air Patrol

Miscellaneous	
Local	*Federal*
Public works departments	Bureau of Mines
Welfare departments	Geological Survey
Flood control districts	Army Corps of Engineers
Cemetery district	Department of Agriculture
Civil defense	Weather Bureau
State	Small Business Administration
Mines or geology departments	Federal Emergency Management
Seismic safety offices	Agency
Civil defense	
Highway departments	

would not apply to federal authorities.) The matter was only resolved after the county attorney was consulted and declared that the sheriff was the responsible authority (Drabek, 1981:35; Kilijanek, 1980:28,32).

EXAMPLE: *Coliseum Explosion, Indianapolis, Indiana, October 31, 1963.* Initially, this disaster was characterized by a general lack of coordination. Contributing to this state of affairs was ambiguity about who should be in charge. The Indianapolis Civil Defense plan specified that the County Civil Defense Director would assume command of all emergency organizations in the event of a major disaster. But in this case, there was some reluctance to call the incident a "major" disaster. City Fire Department officials would normally be in command of a fire within the city limits, but local statute also specified that the County Coroner was the absolute authority in a disaster where a death was involved. The first control was actually

Table 4-4. Private Sector Organizations Involved in Disaster Response

Private hospitals	Private hazardous spill cleanup companies
Physicians, nurses, and allied health professionals	Manufacturing plant fire brigades
Private ambulance companies	Poison control centers
Volunteer search and rescue teams	Explorer Search and Rescue
National Ski Patrol	Private utility companies
Rescue Dog Association	Amateur radio organizations
Red Cross	Veterinarians
Salvation Army	Funeral services
Religious disaster assistance and social organizations	Commercial radio and T.V. stations

Chemical Manufacturers Association—CHEMTREC (hazardous materials telephone hotline)

Railroad, airline, maritime, trucking, pipeline, petroleum, mining, and chemical firms whose equipment or products are involved in a mishap

assumed by the City Police Chief, until 3 hours into the event. At that time, the Superintendent of the State Police raised the point that the Coliseum was located on the State Fairgrounds, and therefore, was under the jurisdiction of the State Police. Responsibility for direction of the disaster operations was transferred accordingly (Drabek, 1968:20,166).

The Private Sector

Responsibility for public welfare and safety in the United States is not limited to governmental bodies, but is also relegated to private sector organizations, businesses, and professionals (FEMA, 1983d:27). Disasters not only cross political boundaries, but also the traditional areas of private sector responsibility. Examples of the types of private organizations and institutions that may become involved are given in Table 4-4.

Non-routine Tasks

Another way in which disasters differ from routine emergencies is that they frequently create non-routine tasks. They also create tasks for which no organization has clear-cut responsibility. Often these tasks have no counterpart in routine emergency operations, and there are no precedents to help decide who is responsible for them (FEMA, 1983d; Quarantelli, 1965:111; Quarantelli, 1982b:159; Bush, 1981; Dynes, 1981:29; Parr, 1970:426; Dynes, 1978:51; Drabek, 1986:29). Examples include:

Situation Analysis

Disasters are characterized by great uncertainty. Often the character and extent of damage and the secondary threats (leaking chemicals, downed power lines, weakened dams) are not immediately apparent and therefore the necessary countermeasures not undertaken. Initial actions are undertaken based on vague and inaccurate information. Disasters are also very "fluid" in nature with needs changing minute-to-minute.

This fluidity necessitates a procedure for determining and updating what the overall disaster situation is and what problems need to be tackled. Typically, it is unclear to the responders who has the responsibility for this task, and in many disasters the process is neglected. When assessment of the disaster situation is carried out, it is generally done independently by a number of individual organizations. Often each agency limits its assessment to those observations of direct consequence to that particular organization. In many cases, the information obtained by these individual organizations is not shared or pooled. Accordingly, an overall picture of the scope, severity, and types of disruption and damage does not emerge early in the crisis (Parr, 1970:425; Golec, 1977:169,174; Quarantelli, 1983:67,91; Quarantelli, 1981a:23; Yutzy, 1969:118,152; Rosow, 1977:72,136,167,193; Drabek, 1986:170,186; Tierney, 1977:154). This failure may result from lack of planning or lack of familiarity with established plans.

Multi-organizational Resource Management

Disasters often create the need for different organizations to share resources (personnel, vehicles, equipment, supplies, and facilities). They also create the need for unusual resources not commonly used in daily emergencies (e.g., search dogs, satellite communications, cranes). Resources in disasters arrive from many atypical sources and often in large numbers. In addition, they are often not dispatched or requested in the routine manner. Units often respond without being asked after hearing of the disaster on their scanners, or on commercial newscasts (Quarantelli, 1983:61; Lewis, 1980:863; 1981f:39; Gordon, 1986:27; Neff, 1977:184; Stallings, 1971:12; Kallsen, 1983:26; Rosow, 1977:105). The atypical mode in which resources respond makes it difficult to tell what resources are present, where they are, what they are doing. Accordingly, it is difficult to determine what resources need to be requested or discouraged from responding. Disasters therefore create the need for procedures aimed at managing and keeping track of resources on a multi-organizational basis. (Further discussion of resource management may be found in Chapter 6.)

Inter-agency Communications

Coordination of multi-organizational task accomplishment, situation analysis, and resource management requires inter-agency communication. The requirement is not only for communications hardware (e.g., radios with compatible frequencies) but also for communications *procedures*. Persons having information need to know who needs it and how to get it there. Persons exchanging information need to use mutually understood terminology. (Inter-agency communications are discussed further in Chapter 5.)

Logistical Support

When organizations respond to a disaster, especially if they come from some distance away and need to stay in the disaster area for an extended period of time, they may require logistical support that cannot be provided in the routine manner (Kallsen, 1983:28; 1983). These needs may include:

— Fuel and maintenance for vehicles

— Sanitary facilities (latrines, showers)

— Food

— Shelter and sleeping facilities

— Relief and replacement personnel

— Emergency message contact arrangements

Figure 4-1.   Disasters often require the establishment of logistical support arrangements such as feeding facilities. (Courtesy of California Department of Forestry, Sacramento, California.)

Search and Rescue

In the typical medical emergency, an ambulance is dispatched to a known location with a definite number of victims. In disasters, however, the situation often requires looking for casualties whose exact number, location, and condition are unknown. This entails the need for widespread organized search and rescue efforts.

Federal guidelines stress the importance of specifying responsibility for search and rescue in disaster planning and operations (FEMA, 1984c:III-2; FEMA, 1985d:3-4; FEMA, 1983e:485). Nevertheless, search and rescue in many disasters has not been perceived as the primary responsibility of any of the participating local government agencies. State statutes have not helped to clarify the responsibilities. A survey published in 1979 was able to find only four states with laws specifying what agency was to be in charge of post-disaster search and rescue (FEMA, 1983b:203). Accordingly, initial disaster search and

Figure 4-2.   In contrast to daily emergencies, disasters often call for large-scale search and rescue operations as in this photo of the San Fernando Valley, California, earthquake on February 9, 1971. (Courtesy of Los Angeles County Fire Department, Los Angeles, California.)

rescue has often occurred in a haphazard manner with little structure or control (Quarantelli, 1983:63; Wenger, 1986:32; Dynes, 1970:432).

Triage and Casualty Distribution

Ambulances responding to a routine emergency usually treat one or two casualties and transport them to a single hospital. In disasters, however, there are usually more patients than one ambulance crew or hospital can handle. Therefore, the need often exists for triage (that is, determining priorities for treatment and transport) and initiating a procedure to distribute casualties equitably among the various hospitals (Gibson, 1977:196; Tierney, 1985a:80; Quarantelli, 1983:63; Barton, 1969:69). (Triage is discussed further in Chapter 8.)

Casualty Lists

Casualty lists are important for two reasons:

— They are needed to address the inquiries of concerned loved ones. After many major disasters there is an inexorable flood of inquiries from concerned loved ones seeking information about the missing (Ross, 1982:64; Worth, 1977:164; Quarantelli, 1983:82).

— They are necessary to determine the number of missing victims for which search and rescue operations must be carried out. This task can be difficult if the missing have no relatives, were out of town when the disaster struck, or were visitors from out of town (Yutzy, 1969:122; Kilijanek, 1981:127).

In many disasters though, a single organization does not have clear-cut responsibility for maintaining casualty lists. Often the task will be attempted by the Red Cross and/or Salvation Army, but other organizations will also become involved (Kilijanek, 1981:78,128; Yutzy, 1969:122). In some disasters the Red Cross attempted to put together casualty lists, but their efforts were stymied by those who were unfamiliar with their function in this regard. Hospitals would not release the information to them for fear of breaching patient confidentiality (KC Health Dept, 1981:24B; Drabek, 1968:76).

Issuance of Passes

An important security task in disasters is keeping unauthorized persons out of the disaster area. This is often done to prevent looting, to decrease congestion hampering rescue efforts, and to prevent persons from being injured in the wreckage. Often this implies the need for passes to let in persons who have legitimate reasons to be there (for example, homeowners and businessmen retrieving salvageable goods and belongings). The lack of precedence may lead to disputes regarding who has the authority to issue passes. In some cases, several organizations may assume the task, resulting in arbitrary and inconsistent enforcement of area restrictions and protests by irate citizens (Rosow, 1977:32; Tierney, 1985b:34; Sorensen, 1981:46; Moore, 1958:17; Quarantelli, 1982b:12; Yutzy, 1969:114).

> **EXAMPLE:** *Earthquake, Anchorage, Alaska, March 27, 1964.* The damaged downtown area was cordoned off, and property owners and businessmen clamored for access to their residences and stores. A "disaster control group" had been organized and began to issue passes to individuals with legitimate reason to enter the area. Anchorage Civil Defense also issued passes to virtually everyone who requested them. To further complicate matters, passes were also being issued by the State Civil Defense, the city building inspector, the police, and by other officials. Some persons with passes were not permitted entry because some of the guards had not been instructed which passes to accept as legitimate (Yutzy, 1969:116).

> **EXAMPLE:** A hospital was flooded after a hurricane when a rain-swollen river overflowed its banks. For security reasons, access to the neighborhood was controlled by roadblocks manned by guards. The recovery operations of the hospital were hampered, because these guards would not honor the identity cards of hospital employees, who were trying to obtain supplies and equipment to clean up and repair the facility. Finally, the workers had to resort to wearing hospital lab coats so the guards would think they were doctors and let them pass (Blanshan, 1978:194).

Hazardous Material Problems

As communities gain more experience with hazardous materials incidents, which are increasing in frequency, the required coordination and technical procedures have become more familiar. In some communities, however, handling of a hazardous spill disaster still fits in the realm of a new task for which smoothly functioning procedures have yet to be developed.

There are cases where a great deal of ambiguity exists as to who has responsibility to plan for and respond to hazardous chemical accidents. For example, accidents in private chemical plants have occurred that were not considered the responsibility of public safety agencies until the hazard extended beyond the plant's boundaries.

There has also been disagreement regarding who was thought to be responsible for handling hazardous material transportation accidents. In one study, a variety of organizational respondents were questioned who should be responsible. Depending on who was asked, the answer was the manufacturer, the transporter, the state environmental protection agency, the state police, the local fire department, the military, or some other organization.

This ambiguity has sometimes been compounded by a tendency for hazardous materials transportation accidents to occur at the entrance from a highway onto private property, or where a public road crosses a private railroad line. The situation has been further complicated when laws designate certain types of local incidents as federal responsibility. For example, hazardous spills involving a navigable waterway have come under the jurisdiction of the U.S. Coast Guard, which has superseded all state and local authority in such cases (Quarantelli, 1981c:33,72,94).

Handling of the Dead

Mass handling of the dead creates problems that may not have been faced in routine emergencies. For example, in hazardous materials accidents, contamination of the body and personal effects sometimes makes routine handling impossible (Dorn, 1986:120). Media attention and the lack of accurate information as to who all the victims are generates inquiries about the dead that can be national in scope (Fritz, 1956:36). Collecting information for such inquiries can be different than in routine fatalities when the disaster crosses jurisdictional boundaries. In a disaster, special materials may be needed for the recovery,

Figure 4-3.   Handling the dead poses different problems in disasters. This photo demonstrates the management of dead bodies after the AeroMexico crash that occurred in Cerritos, California, August 31, 1986. (Courtesy of Los Angeles County Fire Department, Los Angeles, California.)

identification, and care of fatalities. For example, special markers may be needed to indicate where the bodies were found; special body pouches or other containers may be required; and special provisions (for example, refrigerator trucks) may need to be acquired for the storage of large numbers of bodies (Raether, 1986:178).

Other Tasks

Other examples of tasks that may be unique to disasters include:

— Warning and communicating with the public (see Chapter 9)

— Shelter and feeding of displaced persons

— Evacuating neighborhoods

— Evacuating hospitals, prisons, nursing homes, and psychiatric facilities

— Coordinating volunteers (see Chapter 6)

— Acquiring and allocating unusual resources (see Chapter 6)

— Dealing with mass animal carcasses

— Dealing with livestock or family pets that had to be left behind or sheltered (Drabek, 1986:116)

— Procedures for condemning damaged buildings (Moore, 1958:84)

— Disposing of unclaimed valuables and merchandise found in the rubble at the scene (Moore, 1958:85)

— Control of air traffic (Seismic Safety Comm, 1983:15,45,70,75; Drabek, 1981:179)

— Disposing of large amounts of donations (Fritz, 1956) (see Chapter 6)

— Controlling emergency vehicle traffic, so access routes are not blocked by emergency vehicles whose drivers have parked and left them (Hamilton, 1955:50; Drabek, 1968:7,11,19; Cohen, 1982a:102; Morris, 1982)

— Checking on hospitals, nursing homes, and day care centers that may need assistance, but are without communications to call for it (1971:28; Dektar, 1971; Seismic Safety Comm, 1983:91)

— Deciding when and in which areas utilities should be cut off (Seismic Safety Comm, 1983:122)

Figure 4-4.   Management of Livestock from evacuated or affected areas creates unique problems in disasters as in this photo of the "Fourty-Niner Fires" of Nevada County, California in September, 1988. (Courtesy of *The Union*, Grass Valley, California.)

Figure 4-5.   Control of air traffic, especially helicopters, is an increasing concern in disasters as seen at the MGM Grand Hotel fire, November 21, 1980, in Las Vegas, Nevada. (Courtesy of Clark County Fire Department, Las Vegas, Nevada.)

Figure 4-6.   Management of emergency vehicles may be a problem in disasters. This is illustrated by the congestion at the MGM Hotel fire. (Courtesy of Clark County Fire Department, Las Vegas, Nevada.)

Inability to Use Normal Response Tools and Facilities

In addition to being faced with unfamiliar tasks, organizations are sometimes faced with the loss of familiar response tools and facilities. Although damage to hospitals and interruption of their water and power supplies are rare in U.S. disasters (Quarantelli, 1983:81), they are a particular threat in areas of high seismic risk. More commonly, telephones, which are the routine means of communication (especially inter-organizational communication), are unusable. Even when lines have not been damaged, jammed phone circuits prevent normal outgoing communications except from telephones designated as "essential services" (those given special priority by the phone company when trunk lines are overloaded). Another "response tool" which may be, to some extent unusable, is the road and highway system needed to transport disaster response equipment and personnel.

Formation of New Organizations

When the demands of the disaster cannot be met by existing organizations, new organizations may evolve spontaneously to fill the gap. Often, they are very informal in nature and may quickly disband when the immediate crisis is over. Search and rescue, for example, is often carried out by a mixture of citizens, volunteers, and members of emergency agencies who have never

worked together before. Not uncommonly, as they proceed, a transient informal network of coordination will develop. In essence, a temporary, new organization is formed (Dynes, 1974:146; Dynes, 1981:53; Drabek, 1986:218).

> **EXAMPLE:** *Hyatt Hotel Skywalk Collapse, Kansas City, Missouri, July 17, 1981.* "A smoothly functioning, high-performance organization was developed on the spot. Leaders emerged, and were recognized and allowed to lead because they were capable, willing, and because it was necessary. . . . People . . . formed an organization, almost departmentalized, with managers, assistant managers, and a work force (sic) . . . . These 'department heads' worked together almost as though the organization had evolved formally and over a period of years." (Stout, 1981:45)

Figure 4-7.   During rescue operations at the Kansas City Hyatt Regency Hotel skywalk collapse "a smoothly functioning, high-performance organization was developed on the spot." (Courtesy of Kansas City Fire Department, Kansas City, Missouri.)

EXAMPLE: *Tornado, Flint, Michigan, June 8, 1953.* One of the largest contractors in Flint undertook to organize the resources of several big private companies who were donating heavy construction equipment and crews for road clearance and search and rescue. His office functioned as an informal rear headquarters. The contractor's own equipment had two-way radios, and he placed a radio-equipped car in the field to act as a sort of command post. Thus, he set up a working organization and made its services available to the local authorities (Rosow, 1977:143).

One new type of organization that eventually evolves rather typically in disasters is some form of coordinating "committee" or group. The various organizations involved in the disaster response may become aware that their individual and independent activities are inadequate. It becomes clear that a system for inter-organizational information sharing and coordination is necessary. A meeting finally occurs where representatives of the various organizations are invited, and which results in the establishment of some form of multi-organizational coordinating group (Bronson, 1959:42; Dynes, 1978:61; Dynes, 1981:30,42; Drabek, 1986:161,182,186; Kilijanek, 1981:71; Rosow, 1977:20,122, 124; Yutzy, 1969:59,60,77,121). Unfortunately, the formation of such a group may not be accomplished in time to benefit many of the victims (Moore, 1958:15; Kilijanek, 1981:71; Stallings, 1971:25; Rosow, 1977:20, 122, 124; Yutzy, 1969:59,60,77,121; Drabek, 1986:182; Faupel, 1985:35).

EXAMPLE: *Tornado, Jonesboro, Arkansas, May 15, 1968.* Local organizations worked separately for the first 5 hours. As initial search and rescue activities drew to a close, several public safety agencies, along with city and county officials had developed something of an emergency coordination group. Organizations represented at this center included the National Guard, sheriff's office, the state highway patrol, the city police, the mayor, and the county judge. Each group set up its own radio-equipped vehicles outside the police station, which became the emergency operations center. And although direct radio contact among the organizations was not possible, they were close enough to each other for runners to pass information, requests, and instructions among them (Stallings, 1971:25).

EXAMPLE: *Tornado, Waco, Texas, May 11, 1953.* The tornado struck at 4:40 p.m. (Jaworski, 1954:129), but coordination did not even begin to emerge until a meeting at state police headquarters at 11:30 p.m. "We finally organized a disaster commit-

tee, with the power to make the decisions and . . . pass final
judgment on any particular question. . . . " (Moore, 1958:14)

NEW DIVISIONS OF LABOR AND RESOURCES REQUIRE COORDINATION

In disasters, the alterations of traditional divisions of labor and resources
increase the need for multi-organizational and multi-disciplinary coordination
of the various responding participants. Without this coordination, resources
may not be shared or distributed according to need. Disaster-related activities,
such as search and rescue, traffic control, medical care, and transportation of
casualties, may be carried out in a loosely structured, spontaneous manner,
with insufficient communication and control. The result can be duplication of
effort, omission of essential tasks, and even counterproductive activity (Parr,
1970:425; Wenger, 1986:24,26,32,33; Kilijanek, 1981:126).

> **EXAMPLE:** "During a large-scale fire emergency the water
> department issued a call to the citizens to hold the use of water
> to an absolute minimum so that water pressure could be kept
> up for the fire departments. At the same time, however, fire
> officials were on T.V. instructing citizens to wet down their
> roofs with garden hoses." (FEMA, 1981:3)

> **EXAMPLE:** *Volcano Eruption, Mt. St. Helens, Washington, May
> 18, 1980.* Response to this disaster was a large and complex
> undertaking. At least four emergency operations centers and
> five different base camps were a part of the 14-day operation.
> Search and rescue covered 600 square miles, eight to nine times
> over, and involved 2,000 personnel from a multitude of organi-
> zations. At least 100 people were saved and 34 bodies recov-
> ered. It was one of the largest search and rescue missions in
> United States history. Unfortunately, the operations of the
> various organizations were not coordinated. Finally, on the
> third day, representatives from the three county sheriff's de-
> partments and the U.S. Forest Service met and decided to pull
> their operations under a joint decision-making team composed
> of a representative from each of the four agencies. The National
> Guard, however, continued to act independently of this group.
> Lack of inter-agency coordination resulted in several near mid-
> air collisions among the numerous aircraft at the site. It was not

Figure 4-8.   From the 14,000 foot summit of nearby Mt. Adams, climber Vincent Larson captured this photograph of erupting Mt. St. Helens. Fortunately, in spite of being enveloped in ash and fallout within 15 minutes, the climbing party was able to get off the mountain alive. (Courtesy of Vincent R. Larson.)

until the fifth day that the National Guard became integrated into the cooperative effort (Kilijanek, 1981:iii,68,71,74; Drabek, 1981:169).

Evaluations of a number of U.S. disasters illustrating difficulties in coordinating response are summarized in Table 4-5.

The term "mass assault" was used by early researchers to describe the manner in which they observed tasks being carried out at the scene of a disaster. Shortly after impact, there was a massive influx of public safety agencies, equipment, and volunteers. Together with civilians who happened to be in the area, these responders spontaneously came together as informal teams. Under the pressure of great urgency, responders plunged into the first obvious problem they met, wrestled with it until it was overcome by sheer force of numbers, and then moved on to tackle the next problem that confronted them. Little attention was paid to anything except the particular task immediately at hand (Rosow, 1977:16).

> **EXAMPLE:** *Tornado, Flint, Michigan, June 8, 1953.* One of the worst disasters in Michigan history was the tornado that struck the Flint-Beecher area at 8:29 p.m. on June 8, 1953. It destroyed 340 homes and caused major damage to 107 more. It left in its wake 115 dead and 800 injured. The rescue response was fragmented and disorganized. Several emergency response

Table 4-5.   Coordination Problems in Disasters

Disaster	Observations
Tornado Flint-Beecher, MI 1953	The loose control was evident in an uneven distribution of resources in the field (Rosow, 1977:131).
Tornado Waco, TX 1953	It was not until the day after the tornado that a coordinating organization materialized (Moore, 1958:50).
Tornado Worcester, MA 1953	The work of independent agencies was largely uncoordinated (Rosow, 1977:66).
Earthquake Anchorage, AK 1964	Search and rescue was uncoordinated; systematic search of the rubble was not organized until the second day (Yutzy, 1969:149).
Train wreck Chicago, IL 1972	Central control did not exist; a coordinating communications center was not functioning (Cihlar, 1972:17).
Volcano eruption Mt. St. Helens, WA 1980	Real multi-organizational coordination did not begin to take shape until the fifth day (Kilijanek,1981:79).
Hyatt Skywalk collapse Kansas City, MO 1981	There was lack of coordination in obtaining equipment at the scene (Gray,1981:70; Stout,1981:42).
Air Florida crash Washington, DC 1982	There was no single, on-scene commander. Traffic control at the scene was hampered by divided command and lack of central control (Adams,1982:54).
Metrorail crash Washington, DC 1982	There was very little coordination and control (Edelstein,1982:161).
Earthquake Coalinga, CA 1983	Poor coordination among responders resulted in misunderstandings, delays, and duplication of effort (Tierney,1985b:33).

organizations were involved, but they did not coordinate their activities. A member of a Flint Fire Department rescue team described their activity:

"We would be working our way down this block from one house to the next. But there was some other gang ahead of us and another following right behind, maybe 30 feet away, looking through the place that we just finished. We would shove around a pile of timbers and junk to search through underneath and when we'd finish, the team coming afterwards would push it back to check where we had dumped it."

When asked if this was the same pile of junk that the team ahead of him had shoved around, the firefighter admitted that indeed it was. Nobody checked on his team's work, nor did the team report to anyone the results of their work. There was a multitude of search teams at work, but none knew what the other was doing, and no one was trying to keep track (Rosow, 1977:130).

In spite of improvements in disaster coordination since this classical example from the Flint tornado, one still can observe multiple organizations operating independently without knowledge about what other organizations involved in the disaster response are doing. For example, in a 1986 Disaster Research Center study of six disasters, major problems with coordination occurred in four of them (Wenger, 1986:23,44).

The management of many emergency response and public safety agencies is patterned on the **military model.** This reflects the belief that the most effective emergency operations are carried out under rigid control exercised from a single commander. Indeed, such a centralized intra-organizational authority structure may be entirely appropriate and effective in the independent, daily, routine operations of these organizations.

In the United States, however, no single organization can legitimately control what all other public and private organizations do and don't do in a peacetime disaster (Drabek, 1980:23; Drabek, 1981:xx; Drabek, 1985b:9; Drabek, 1987:106; Dynes, 1981:29). It has been argued, therefore, that realistic disaster management in a country with a decentralized government such as the United States, with its traditional preferences for local control and private enterprise, probably cannot be accomplished using a military model. Rather, coordination among the various independent responding organizations needs to be based on negotiation and cooperation (Drabek, 1980:23; Drabek, 1981:122; Drabek, 1987: 92,239; Kilijanek, 1981:126; Adams, 1981b:2,52,61; Dynes, 1981:29).

Although it may not be obvious initially, the need for joint decision-making eventually becomes apparent in most large disasters.

The need for joint, inter-organizational direction and decision-making is reflected in three processes which are being used with increasing frequency in disaster responses. These are **multi-agency pre-disaster planning** (see Chapter 3), **emergency operations centers** (see Chapter 6), and the **unified command structure** of the Incident Command System (see Chapter 7).

> **PRINCIPLE**
> In contrast to most routine emergencies, disasters introduce the need for *multi-organizational* and *multi-disciplinary* coordination.

SUMMARY

Disasters may generate a whole host of problems that are not found in routine emergencies. Organizations change structure, with various positions being filled by different persons. Multiple organizations are faced with overlapping areas of responsibility. Many activities are taken on by unsolicited volunteers. New tasks, sometimes requiring unusual resources, present themselves for which no one has clear-cut responsibility. New organizations even come into being. Multiple organizations are faced with the need to coordinate activities with each other on a moment-by-moment basis, without familiar procedures for carrying this out. Furthermore, all of this may take place under conditions of extreme urgency, which virtually precludes the time required to develop the necessary coordination.

PLANNING CHECKPOINTS

[ ] Does your disaster plan include procedures for determining responsibility for disaster tasks that are not the traditional responsibility of any single organization (for example, overall situation assessment, search and rescue, casualty distribution)? For tasks for which multiple organizations may claim responsibility?

[ ] Does the plan include provisions and procedures for a multi-organizational coordination body?

[ ] Does the plan make provisions to incorporate responding public and private organizations that do not usually play a part in routine emergencies?

5

INTER-AGENCY COMMUNICATIONS

Adequate communication is a recurring challenge in disaster response. (Courtesy of California Office of Emergency Services, Sacramento, California.)

In disasters, communication difficulties are often hard to separate from coordination difficulties, and the greatest coordination difficulties are inter-organizational. Therefore, many of the communications problems are those related to inter-agency information sharing. Frequently, the means for communication exists, but for a number of reasons, persons are hesitant to communicate with others outside their own organization.

Inter-organizational communication is fostered by those factors which promote trust in other organizations and familiarity with how they function. These include: informal contacts, joint planning and training, preplanned agreements for the division of disaster responsibilities, and the use of similar terminology, procedures, and performance criteria. Inter-organizational radio networks, common mapping systems, and computer networks also contribute to effective communications.

79

COMMUNICATIONS PROBLEMS IN DISASTERS

One of the most consistent observations about disasters is that communication is inadequate. An in-depth 1986 study suggests that this is a continuing problem. Major communications problems were found in half of the six disasters evaluated (Wenger, 1986:11,14,44,76). Less clear is what constitutes adequate communication, and why it is so difficult to achieve.

In this chapter, the discussion of inter-organizational communication will focus on two main topics: 1) "pre-incident" communication; and 2) technical aspects of communication. Inter-organizational communication as it relates to resource management is discussed in Chapter 6.

RELATIONSHIP OF COMMUNICATION TO COORDINATION

Types of Information Needed for Coordination

The importance of communication is its ability to get people to work together on a common task or toward a common goal—to coordinate. It is the process by which each person understands how his individual efforts intermesh with those of others. Frequently, what are perceived as communications problems are actually coordination problems in disguise (Brunacini, 1985b:54). Disasters pose unusual demands for *inter-organizational* coordination. For this reason, a substantial portion of disaster communications problems are related to the exchange of information among organizations. The most crucial types of information that need to be shared are those related to:

— an ongoing assessment of what the disaster situation is and what disaster countermeasures need to be undertaken;

— an ongoing determination of what resources are needed to undertake the countermeasures, what resources are present, assigned, out-of-service, what resources are available, how they can be obtained, what is their capacity, and how long will it take for them to arrive;

— a determination of the priority of needed disaster countermeasures (and, therefore, resource allocation); and

— a determination of what persons and organizations will be responsible for the various tasks necessary to accomplish the countermeasures (Sorensen, 1985:32).

"People" Problems in Communication

Research on disasters suggests that many communications problems are *"people problems,"* rather than *"equipment problems"* (Kilijanek, 1979:7; Quarantelli, 1965:109; Quarantelli, 1985:12; Worth, 1977:160). Communication equipment may be in short supply, but more often than not a physical means of communication is available (Quarantelli, 1985:12). Examples of "people" problems in communication include the following (Kilijanek, 1979:5; Rosow, 1977:139; Drabek, 1986:54; Quarantelli, 1983:106):

- The "Robinson Crusoe syndrome," or "we're the only ones on this island." Organizations are accustomed to operating autonomously and fail to change this approach in disasters where multiple organizations are involved and are dependent on one another. Each person gives priority to the information needs of his own organization rather than that of the overall response effort.

- Terminology and procedures used to exchange information vary among different organizations.

- There is a hesitancy to depend on other organizations, often due to lack of trust or familiarity, or due to political, jurisdictional, and personal disputes.

- There is no mutual agreement as to who has the responsibility for the collection and dissemination of various types of information, or to whom it should be distributed.

- Persons possessing information do not realize that another person who needs it, doesn't have it.

- The information needs of other organizations are not understood.

These are crucial aspects of communication that no amount of radio equipment is likely to correct (Rosow, 1977:173).

PRINCIPLE

In disasters, what are thought to be "communications problems" are often coordination problems in disguise.

THE IMPORTANCE OF "PRE-INCIDENT" COMMUNICATIONS

One key to understanding disaster communication problems is the concept of pre-incident communications. In efficient *routine* emergency operations, the vast majority of communications have occurred *prior to the incident*. The goals and tasks are often determined by tradition. They are formalized in statutes, contracts, and charters. Within various organizations, they are addressed in rules, regulations, performance standards, and standard operating procedures. The important point here is that many of these tasks are *known beforehand* and do not have to be communicated for each event to which an organization responds (Dynes, 1978:59; Dynes, 1981:39).

Yet, a number of observations in *disasters* have revealed a lack of pre-impact communications among key local disaster response organizations such as law enforcement agencies, fire departments, local emergency management agencies, and organizations in the health and welfare sectors (Wenger, 1986:76; Quarantelli, 1978:4). This may be compounded by the fact that organizations responding to disasters often include many who have had minimal previous contact, because they do not respond to local emergencies on a routine basis (Quarantelli, 1983:64; Kilijanek, 1981:50).

When organizations have interacted and coordinated with each other *beforehand*, they have had fewer problems doing so in a disaster (Kilijanek, 1981:50,126; Dynes, 1978:58; Adams, 1981b:40,53; Tierney, 1977:155; Dynes, 1977:12; Drabek, 1986:125; Sorensen, 1985:32).

> **EXAMPLE:** *Hyatt Hotel Skywalk Collapse, Kansas City, Missouri, July 17, 1981.* Although Kansas City's ambulance crews all work for a private ambulance company, they are quartered at city fire stations. It was felt that because the ambulance and fire personnel work together daily and share the same facilities, this contributed to the ease with which they worked together during the disaster (Stout, 1981:36).

> **EXAMPLE:** *Tornado, Wichita Falls, Texas, April 10, 1979.* Because of frequent pre-disaster contacts, various agency heads knew each other quite well. This formed a basis for mutual expectations and understanding which facilitated the multi-organizational response (Adams, 1981b:40).

The importance of pre-disaster contact helps to explain a seemingly paradoxical observation made in a number of disasters, that smaller communities with fewer resources tended to coordinate their disaster responses better than larger, more urban areas. The explanation was that, because the emergency response agencies of the smaller communities had fewer resources, they had to coordinate and cooperate with each other to handle even the more routine emergencies. Therefore, when they had to work together in a disaster, it was not the first time they had done so (Quarantelli, 1983:105).

PRINCIPLE
Those who work together well on a daily basis tend to work together well in disasters.

Development of Trust

Even under the pressure of a disaster, certain preliminary information has to be exchanged before meaningful communication and coordination can take place with a member of an unfamiliar organization. Examples of the types of critical information needed include:

— what the organization's legitimate role is in the disaster response;

— whether that person has a legitimate position in that organization; and

— the competence and reliability of that person.

Emergency organizations with disaster operations responsibility frequently hesitate to coordinate with others unless these questions have been addressed. *This hesitancy may exist even though there are formal plans or arrangements for the different organizations to coordinate.* Unfortunately, the urgency of the disaster situation often precludes the time necessary to determine the answers to these questions on-the-spot. The result is that, unless they have been addressed *before the disaster,* there is a reluctance to depend on the activities of other organizations and a failure to coordinate and communicate with them (Dynes, 1978:58; Rosow, 1977:63,74,76).

When one is dependent on other team members, particularly in life-threatening situations, he needs to feel confident in their competence and reliability. Developing this level of trust often requires "pre-incident" contact over a period of time.

Resolution of Political, Personal, and Jurisdictional Disputes

Cooperation is adversely affected by pre-existing personal, political, and jurisdictional disputes. Quarantelli found, for example, that conflicts within the emergency medical services (EMS) system and between it and other community sectors was a major factor hindering disaster planning. Such differences as those between city and county, or public and private interests had a pervasive negative influence on cooperation. The consequences of such disputes may range from the exclusion of organizations from planning meetings to charges that an organization is transgressing another's jurisdiction or responsibility. Unfortunately, jurisdictional disputes unresolved on an everyday basis, do not tend to get resolved in disasters (Quarantelli, 1983:105).

Knowledge of How Other Organizations Function

Inter-organizational teamwork requires information sharing. Persons need to know when they possess critical information required by someone in another organization, how to get it to the other person, and how to use terminology the other person will understand (Wenger, 1986:15,46). For these reasons, knowledge about how other organizations function tends to promote inter-organizational communication and coordination (Dynes, 1978:55,58).

That is why the most effective and cooperative relations seem to develop between similar organizations where each has knowledge of the internal operations of the other. With such familiarity, each organization is more likely to feel it can coordinate and exchange information with the confidence that the other is reliable and competent. Thus, police agencies tend to interact best with other law enforcement groups and fire departments to cooperate best with other fire fighting agencies (Mileti, 1975:80; Dynes, 1978:58; Rosow, 1977:185,187). Knowledge about how dissimilar organizations function is more apt to be lacking (Seismic Safety Comm, 1979:36; Quarantelli, 1983:103). Important knowledge about other organizations includes that about roles, resources, needs, terminology, and competence (Wenger, 1986:33).

Knowledge About Routine Function

In some cases, at least to a degree, knowledge of how other organizations *routinely* function is useful in disaster situations. For example, familiarity with another organization's terminology or competence, engendered by previous contact during routine emergencies, is likely to facilitate interaction during a disaster.

Knowledge About Pre-planned Disaster Functions

In contrast, knowledge about other organizations' disaster response needs and roles are not always reliably based on familiarity with their routine functions.

The responsibility for a number of important disaster roles (especially those that have little counterpart in routine emergency responses) may be ambiguous unless there is a functioning, pre-disaster, mutual agreement that clarifies the issue. In that case, knowledge about the agreement can convey knowledge about how other organizations will function. For the purpose of formulating such agreements, federal disaster planning guidelines suggest bringing together representatives of the various organizations likely to be involved in a disaster response. These agreements may be summarized in the form of disaster role matrices. A number of role matrices may have to be developed depending on the type of disaster and the jurisdictions it involves. An example of such a role matrix for a county-level disaster is shown in Fig. 5-1.

Certain organizations in disasters have needs that they must rely on other organizations to fulfill. For example, in order to muster their resources, hospitals need to have advance warning that they will be receiving patients and timely estimates of the types, numbers, and severities of casualties to be expected. This information must come from those at the disaster scene. In addition, hospitals are at the mercy of those at the scene to see that casualties are equitably distributed, so that no one hospital receives an inordinate number. As discussed later in Chapter 8, the failure to recognize these needs has been the cause for problems in many disasters.

Knowledge about the special disaster resources to which another organization has access can also be helpful. For example, local branches of state organizations (e.g., state police, state forestry) may be able to obtain state resources without the necessity of a state-wide disaster declaration. Certain organizations may have access to unusual resources valuable for disaster operations (e.g., satellite communications equipment, search dogs, devices for chemical analysis).

Standardization

Standardization can also create familiarity with other organizations. To the extent that organizations agree to use the same procedures, resources, terminology, and performance criteria, they will share a common knowledge of each other's function.

The effectiveness of disaster response may depend on the ability of organizations to share resources. However, when requesting resources from another organization, the lack of standardized terminology may make it difficult to know what one will receive. The ability of a requested fire truck to carry out the mission for which it is requested may depend on the equipment it carries, its water capacity, and the number and training of its crew. All of these factors may vary among different fire departments. This problem is now being addressed in areas where the Incident Command System is in use, because it stipulates standardized terminology to describe common emergency response resources (see Chapter 7).

ASSIGNMENT OF RESPONSIBILITIES

P—Denotes Primary Responsibility
S—Denotes Support Responsibility

POSITION	Emergency Operations	Communications	Warning	Evacuation	Shelter Operations	RADEF	Law Enforcement	Fire Services	Health & Medical Services	Search & Rescue	Damage Assessment & Recovery	Resource Management	Social Services	Public Information	Staff Training
Executive Group	P	P	P	P	S	S	S	S	S	S	S	S	S	S	S
Sheriff & Gulfport Police Dept.	S	S	S	S		S	P	S	S	S	S	S		S	S
Gulfport Fire & County Fire	S	S	S	S	S	S	S	P	S	S		S		S	
Damage Assessment	S				S	S					P		S		
Engineering	S	S			S	S			S	S	P	S		S	
Utilities	S	S			S	S					S	S		S	
Plotting	S	S	S	S	S	S	S	S	S	S	S	S		S	
Public Information	S	S	S	S	S	S	S	S	S	S	S	S	S	P	
Rescue	S	S		S		S		S		P	S	S		S	S
Public Works	S	S		S		S			S	P	S		S		
Human Resources				S								S	S		
Military	S	S	S	S	S	S	S	S	S	S	S	S		S	
State/Federal	S	S	S	S		S	S	S	S	S	S	S	S		S
Shelter	S	S	S	S	P	S					S		S	S	
Message Coordinator	S	S											S		
Logistics	S	S	S	S	S	S	S	S	S	S	S	P	S	S	S
Director	P	P	P	P	S	S	S	S	S	S	S	S	S	P	P
Deputy Director	S	S	S	S	S	S	S	S	S	S	S	S	S	S	S
Administrative Officer	S	S	S	S	S	S	S	S	S	S	S	S	S	S	S
RADEF Officer	S	S	S	S	S	P	S	S	S	S	S	S	S	S	S
Medical Officer	S	S		S	S	S	S	S	P	S	S	S	S	S	S
Welfare Officer	S			S				S			S	S	P	S	S
Support Staff	S	S	S	S	S	S	S	S	S	S	S	S	S	S	S
Communications Officer	S	P	P	S	S	S	S	S	S	S	S	S	S	S	S
Communications Assignees	S	S	S	S	S	S	S	S	S	S	S	S	S	S	S

Figure 5-1.   Role matrix—Assignment of Responsibilities.
(From Emergency planning: student manual, SM-61, Washington, D.C., 1983, Federal Emergency Management Agency.)

Many police departments, fire departments, rescue units, and ambulance services use radio codes for communication. These codes may vary from agency to agency, a situation which interferes with inter-organizational communication.

> **EXAMPLE:** The California Highway Patrol uses the code "11-79" to dispatch a patrol car to an accident and to advise that unit that an ambulance is en route. One of the local sheriff's departments uses the code "901T" to mean the same thing.

For this reason an increasing number of agencies are abandoning the use of radio codes in favor of simple English ("Clear Text" or "Clear Speech") (Publ Safety Dept, 1975:2; ICS, 1982:26).

The likelihood that one organization will interact with and depend on another organization is enhanced when it is perceived that its members are competent. This is facilitated when members of such groups pass a standardized test of their knowledge, skills, and competence. Some departments require a rigorous qualifications exam as a condition of employment, which may include a test of basic knowledge as well as a test of physical agility and endurance. More advanced examinations may be required at the completion of a rookie's training and further tests become necessary for promotion. In some locales, certain public safety occupations require passing standard state exams (e.g., state standards for peace officer, firefighter, or paramedic training). In some cases, national qualification standards exist. Examples include the National Registry of Emergency Medical Technicians, the certifying examination given by the American Board of Emergency Medicine, and the qualifications for Incident Command System training under the National Interagency Incident Management System.

Joint Planning and Training

One of the most important ways in which disaster response organization members can get to know and trust each other and become familiar with the function of other organizations is during joint planning and training activities (Dynes, 1978:58; FEMA, 1983d:14; Grollmes, 1985:8; Hildebrand, 1980:4; Stevenson, 1981:57; Adams, 1981b:47).

> **EXAMPLE:** *The Rocky Mountain Fire Academy, Denver, Colorado.* The Aurora and Denver Fire Departments began sharing training resources in 1985 to combat the effects of severe budget restrictions and decreased tax revenues. This agreement has catalyzed dialogue not only between the two fire depart-

ments, but among other agencies and municipalities as well. For example, the personnel and administrators from the two fire departments have found themselves comparing standard operating procedures and managerial techniques and adopting the best ones from each agency. As a result of this experience, the feasibility of mutual hazardous materials response and firefighting operations are being investigated as well as the purchase of a new 800 mHz fire dispatching system (Schumacher, 1988:16).

EXAMPLE: *MGM Grand Hotel fire, Las Vegas, Nevada, November 21, 1980.* The county fire department had conducted training sessions to test its plans, as well as annual drills involving civil defense, police, city and county fire departments, the airport, and the local medical community. When the disaster occurred, the years of planning and rehearsal paid off. The fire, despite its magnitude, did not overwhelm local resources (Buerk, 1982:641; Parrish, 1981:12).

Importance of Informal Contacts

Even informal contacts are an important part of "pre-incident" communications (Drabek, 1986:44,125; Quarantelli, 1983:120,130; Dynes, 1977; FEMA, 1983d:205; Sorensen, 1985; Grollmes, 1985:8; Wenger, 1986:33). The value of such contacts is not only the knowledge they generate about how other organizations function, but also in the trust that may develop (Drabek, 1980:10). Even in the absence of well-developed plans, personal familiarity can facilitate coordination (Dynes, 1978:54).

EXAMPLE: *Tornado and Showboat Capsizing, Lake Pomona, Kansas, June 17, 1978.* The responding organizations did not have a well-developed, written plan for responding to a disaster of this type. However, many of them had functioned jointly before in routine emergencies, and what they lacked in formal plans, they made up for because of close personal acquaintances among those in charge of each organization.

"They knew each other well and they knew the local area. They were able to obtain needed resources quickly. Although they did not have a plan specifying who should do what, they put their heads together, considered options, and made decisions which respective units could act on quickly. To no small degree, the success of this operation reflected the sheriff's knowledge of the various organizations and their managers, rather than planning or prior experience. Too often this type of resource is not recognized in emergency management, nor are those responsible encouraged to nurture it. Piles of paper plans stored away neatly on office shelves are poor substitutes for strong interpersonal relationships rooted in trust and weekly contacts." (Drabek, 1981:52,54)

Sometimes communications and resource-sharing between organizations is facilitated when a person has membership in multiple organizations. An example is the police chief who is also a member of the local Red Cross disaster committee, a civil defense group, and a municipal administrative council (Dynes, 1978:55; Sorensen, 1985:32).

> **EXAMPLE:** *The MGM Grand Hotel Fire, Las Vegas, Nevada, November 21, 1980.* Bob Forbuss, the Mercy Ambulance triage officer at the scene, was also a school district trustee. He contacted the school district and arranged the dispatch of 50 school buses to the scene for the transport of survivors to a county evacuation center (Munninger, 1981:38).

There is another type of informal contact which may facilitate communication and coordination among organizations in disasters. This occurs when someone has extensive, long-term friendships with members of other important community organizations. These contacts, and the trust and loyalties which they have fostered, can expedite cooperation among the organizations to which these contacts belong (Quarantelli, 1982c:69; Bush, 1981:4; Orr, 1982:10).

Pre-planned Procedures for Developing a Strategy at the Incident

Although certain aspects of disaster response can be anticipated and planned in advance, each disaster is to some extent unique. Therefore, a specific strategy, or "action plan" has to be created at the time the individual incident occurs. By this is meant an overall, multi-organizational plan. The failure to do this is one of the reasons that different organizations responding to the disaster often do not carry out a unified, concerted effort. This post-impact creation of an incident "action plan" is facilitated if there is a *pre-arranged*, mutually agreeable procedure for the process.

One example of such a procedure which has achieved national acceptance is that used by the Incident Command System (ICS) (see Chapter 7). A simplified description is given of the procedure as it would be used in a multi-jurisdictional disaster incident:

— A survey is made of the disaster to assess what problems need to be tackled (what objectives, or "goals," need to be accomplished) and what resources are present.

— Each agency indicates the problems (objectives) for which it is responsible, that is, "This is what my organization needs to accomplish." (The pre-agreement as to who has what responsibilities may have been formalized in a role matrix like that in Fig. 5-1.) Each agency also indicates the constraints (e.g., fiscal, legal, political) that limit what it can do. The various objectives might be quite diverse as in the following hypothetical example:

Agency A: "I need to evacuate the people from area X."

Agency B: "I must control traffic and prevent the entry of unauthorized persons."

Agency C: "I must provide information to the area hospitals and make triage, field medical care, and casualty transportation arrangements."

Agency D: "I must locate and put out any fires."

— A mutually agreeable multi-organizational strategy is developed for the incident that best meets the objectives and priorities of the various agencies and the overall situation.

— The incident is subdivided into geographic areas and/or general functional areas, and organizational groups assigned to these areas are given specific work assignments which will contribute toward attainment of the incident objectives. Each group indicates the resources it needs to accomplish its work assignment.

— The resources needed to accomplish the work assignments are compared with the resources present at the incident. By this means it is possible to determine what additional resources must be requested.

This ICS incident action planning procedure is not a "committee process" that must somehow resolve all differences in agency objectives before any action begins. It is, however, a "team process," which by means of a sharing of objectives and priorities, formulates a set of collective directions to address the needs of the entire incident and which reduces duplication and omission of crucial tasks. Experience with the system has shown that this collective sharing of information and objectives has led to a voluntary sharing of resources and modification of individual agency objectives to meet the overall requirements of the incident (Irwin:5; ICS, 1985b).

TECHNICAL ASPECTS

Although "people problems" appear to be among the greatest obstacles to effective disaster communications, there are a number of technical problems that can also inhibit information exchange.

Loss of Function

Flooding, tornadoes, hurricanes, and earthquakes are all capable of toppling antennas and interrupting normal electrical power. Commercial broadcast stations and public safety radio networks may be rendered inoperable (Seismic Safety Comm, 1983:15).

> **EXAMPLE:** A recent study in California revealed that a number of the commercial broadcast stations, which were to be used in disasters for state officials to communicate with the public, had no provision for back-up power (JCFEDS, 1983:31,ii-48).

> **EXAMPLE:** *Earthquake, Coalinga, California, May 2, 1983.* During the shaking, the police department's radio console fell on the floor and broke, disabling the system. The earthquake also cut off power to the hospital's radio, and the emergency back-up power system failed (Tierney, 1985b:33; Seismic Safety Comm, 1983:15,16,32).

Effective disaster preparedness requires that essential communications equipment have sources of back-up power. The equipment and reserve power source need to be protected against the forces of the impact. In areas of seismic risk, this equipment needs to be anchored. In flood-prone regions, it needs to be placed in elevated areas. In tornado and hurricane areas, the antennas must be able to withstand the winds, and/or back-up provisions made.

Inter-agency Radio Networks

The Problem of Frequency Incompatibility

Besides loss of function, there are other technical problems in inter-organizational disaster communications. Because of the unreliability of telephone communications in disasters, inter-organizational communications are best carried out by two-way radio. Unfortunately, public safety radio frequencies have been assigned in such a manner as to make this very difficult. Part of the problem is that radio traffic occurs on a number of different "bands." A "band" is a collection of neighboring frequencies, and it is technically possible to have a single radio that can switch to different radio frequencies on the same "band." However, the difference in frequencies on separate bands is so great that completely different radio-electronic circuits and antennas are needed. *In effect, for each band a completely different radio is needed.* If several organizations are on different radio frequencies in the same band, it is possible for them to communicate with each other if they all switch their radios to the same frequency. This is not possible if the frequencies used by the organizations are on different bands (JCFEDS, 1983:18). Unfortunately, this is the way radio frequencies have been assigned. The bands involved are: low band (37–42 mHz); high band (150–155 mHz); UHF (450–470 mHz); UHF-TV (450–470 mHz); and the 800 mHz band (806–902 mHz). In addition there are different bands for use by the military and ham operators. Even units of similar organizations in adjacent jurisdictions cannot talk to each other using their assigned frequencies. Table 5-1 illustrates the prevalence of the problem.

This problem was discussed in a 1983 report by the California State Legislative Joint Committee on Fire, Police, Emergency and Disaster Services entitled *California's Emergency Communications Crisis* (JCFEDS, 1983). The report states that units of the Los Angeles County Sheriff cannot communicate directly with Los Angeles City Police units because they are on different bands. Even the designated law enforcement and fire service "mutual aid" frequencies in California cannot always be used to solve the problem because many of the agencies have been assigned primary frequencies on a different band than the mutual aid frequency. In order to take advantage of the mutual aid frequency, they would have to buy a whole second set of radios, an expense that few jurisdictions are capable of bearing. At the time of the report, only 30% of the law enforcement agencies in the state had vehicles equipped to use the California Law Enforcement Mutual Aid Radio System (CLEMARS) frequencies (JCFEDS, 1983:14,ii-3). The problem is illustrated by the following:

> **EXAMPLE:** *The Norco Bank Robbery and Lytle Creek Shootout, Riverside County, California, May 9, 1980.* Riverside law officers (whose radios communicate on 450 mHz) were chasing bank robbers into San Bernardino County (which uses 150 mHz), and called for help from Los Angeles County (39 mHz). There were 200 fatigue-dressed (SWAT team) law enforcement officers in the hills above San Bernardino, carrying AR-16 rifles, looking for six robbers with fatigues and AR-16 rifles. Because their frequencies were incompatible, officers from the different counties could not communicate by radio, and they had to resort to hand signals and shouting. A sheriff's deputy from

Table 5-1.   Lack of Inter-organizational Radio Networks

Disaster	Observation
Wildland fire Bel Air, CA 1961	Mutual aid fire companies from 23 nearby cities could not be effectively controlled, because they were all operating on different radio frequencies (Bahme, 1978:70).
Hurricane Camille 1969	Lack of a common radio frequency complicated National Guard activities; various pieces of mobile equipment operated on different radio frequencies (Stallings, 1971:13).
Earthquake San Fernando Valley, CA 1971	One conspicuous weakness of the hospital radio network was its lack of tie-in to police and fire radio systems (1971:28).
Tornado Xenia, OH 1974	Lessons: communications can be a problem if there are no common radio frequencies; coordination is difficult if you cannot talk with all field units (Troeger, 1976).
Ammonia spill Houston, TX 1976	A hard look should be given to communications from the hospital to other governmental units (Swint, 1976:45).
Flash flood Bandera, Kendall, & Kerr Co.'s, TX 1978	Inter-agency communications was mentioned as a leading cause of coordination difficulties (Drabek, 1981:89).
Nuclear accident Three Mile Island, PA 1979	Hospital communications must be linked to emergency operations centers (Maxwell, 1982:299).
High-rise fire MGM Hotel Las Vegas, NV 1980	Police, ambulances, public works, fire department, and helicopters were unable to communicate through a central command center at the scene, so many efforts were delayed and coordination difficulties were magnified (White, 1981:31).

cont'd.

Table 5-1.   Lack of Inter-organizational Radio Networks *cont'd.*

Disaster	Observation
Skywalk collapse Hyatt Hotel Kansas City, KS 1981	A coordinated, interfacing communications network should be available to all responding emergency service units (Grollmes, 1985:10; KC Health Dept, 1981:24b).
Aircrash Kenner, LA 1982	Police, fire, and ambulance agencies could not communicate directly with one another by radio (Morris, 1982:63).
Earthquake Coalinga, CA 1983	Major problems included the lack of common radio frequencies for fire, law enforcement, and public utility crews (Seismic Safety Comm, 1983:96).
Tornado Lenoir Co., NC 1984	Overtaxed radio communications were made worse by the lack of a common frequency (Deats, 1985:52).
Aircrash Dallas, TX 1985	Because departments at the scene did not have a common radio frequency, messages had to be conveyed by runners or shouting (Gallagher, 1985:44).
Flood/levee break Yuba County, CA 1986	A common radio frequency is needed for responders who must coordinate with one another (Sac Fire Com, 1986).

Riverside County was killed when the San Bernardino County Sheriff's helicopter saw him driving into an ambush, but could not warn him (JCFEDS, 1983:ii-98; Irwin, 1987).

EXAMPLE: *Flash Floods, Bandera, Kendall, and Kerr Counties, Texas, August 1–4, 1978.* "One county sheriff recounted the intense frustration he felt while standing aside his cruiser that was parked at a ravaging riverbank. He had no way to communicate with a pilot in a helicopter that was hovering over a victim hidden from [the pilot's] view by tree branches. A communication chain linked him to a dispatcher, who, in turn, could reach a state agency that could relay messages to the federal military base that had radio contact with the pilot. The chain was activated minimally, however. When used episodically, these chains reflected the consequences of distortion typically found in such multiperson relays." (Drabek, 1985a:88)

As stated by one emergency services coordinator in California, "Law enforcement agencies cannot communicate with the fire services; the fire service cannot communicate with the California Highway Patrol; the CHP is unable to communicate with units of the Sheriff's Department; cities are unable to communicate with the county . . . , and the list goes on and on." (JCFEDS, 1983:16)

EXAMPLE: *Chlorine Gas Leak, Santa Rosa, California, July, 1982.* This incident required the evacuation of approximately 2,000 people and the response of over 20 different agencies, including four law enforcement agencies, four fire departments, seven county agencies, five state agencies, and three volunteer agencies. During the response, the two cities involved—Santa Rosa and Sebastopol—could only communicate with each other by phone, and the on-scene incident commander could not communicate directly with all of the responding agencies to coordinate their activities. Command, control, and coordination could not be properly exercised, nor could vital information be readily exchanged between two local governments (JCFEDS, 1983:16).

The situation was described as even worse for emergency medical services. As in the case of law enforcement, different neighboring ambulances and EMS units have been assigned radio frequencies on different bands. But, unlike law enforcement and fire services, not even one of these bands had a frequency set aside for mutual aid communications. Because of this, ambulances were reported to lack any effective contact with hospitals or others when transporting over any significant distances. In addition, ambulances had to share the same frequency with non-emergency users such as veterinarians, tow trucks, washing machine repairmen, concrete companies, trucking companies, and boat repairmen. The report criticized the lack of a dedicated radio link between the medical community, the county emergency operations centers, the State Office of Emergency Services, and the State Department of Health Services (JCFEDS, 1983:15,26,98,ii-138).

Lack of radio frequency compatibility likewise was said to prevent communications with other agencies likely to be involved in disaster responses, such as the California Department of Transportation, the county health agencies, and the National Guard (JCFEDS, 1983:ii-19).

Clearly, these emergency communications problems are not unique to California. Since the time of the report, California has taken several steps to correct their problem. The California State Office of Emergency Services has developed an emergency, multi-organizational VHF mobile coordination frequency, CALCORD. This will allow mobile units of fire, law enforcement, medical, and other disaster responding agencies to communicate with each other. In addition, the California State EMS Authority is trying to negotiate a statewide emergency medical services coordination frequency. The EMS Authority has also developed several other sophisticated communications networks (including the use of amateur radio, cellular telephone, computer links) and has state administration approval for satellite communications capability (Koehler, 1987).

PRINCIPLE

Disasters create the need for coordination among fire departments, law enforcement agencies, hospitals, ambulances, military units, utility crews, and other organizations. This requires inter-agency communication networks utilizing compatible radio frequencies.

It would be an oversimplification to suggest that the mere development of a mutual, multi-disciplinary, multi-agency communications frequency will solve inter-organizational disaster communications problems. In fact, the undisciplined use of such a frequency could quickly jam it with excessive traffic, rendering it useless.

> **EXAMPLE:** *Air Florida Crash, Potomac River, Washington, DC, January 13, 1982.* A U.S. Park Police helicopter was the only means by which any of the plane crash victims could be rescued. Even though the aircraft had radio capability to communicate with ground rescue units, it was not able to do so. This was because the frequency was too jammed with emergency and non-emergency radio traffic (Grollmes, 1985:12).

Communication has to be organized on a series of networks in order to accommodate the volume of traffic and direct it in a manner consistent with the inter-organizational management structure. In addition, procedures are necessary to determine channels and priorities for information flow.

Frequency Sharing Agreements

One approach to the problem of inter-agency communications is the development of frequency sharing agreements and radio caches (FEMA, 1987:38,40; ICS, 1980). A frequency sharing agreement is a mutual pact allowing multiple organizations involved in joint emergency operations to share radio frequencies licensed under single organizations.

Technical advances in radio equipment have made frequency sharing easier. Synthesized and programmable radios, some of which are capable of using hundreds of different frequencies, are now widely available. Unfortunately, commercial public safety radios are still capable of transmitting and receiving on only one *band*.

Radio Caches

To some extent, the difficulty of communications among organizations whose radios are on different bands can be overcome by the development of "radio caches." Each radio cache has a collection of multi-channel portable radios each with the same frequencies, extra batteries, battery chargers, and, perhaps, portable repeaters. These caches are stored where they can be obtained and mobilized to provide linking communications nets in multi-agency emergency operations and disasters.

"Two-way" Use of Scanners

One method of communication among agencies on different bands involves the use of modern, programmable scanning *receivers* (receivers that scan a number of frequencies, pausing to listen on each frequency only when it is carrying radio traffic). Although these devices cannot transmit, they can receive on multiple radio *bands*. With this technique, one person transmits on his two-way radio. The message is received on the second person's scanner. The second person then transmits his response on his two-way radio, and it is received on the scanner of the first person (see Fig. 5-2).

This technique can even be used if one person is transmitting on one band (e.g., UHF) and the other person is transmitting on another (e.g., VHF-Low). Coordination frequencies can be determined for each band. Each agency sets up its radios so they can transmit on the coordination frequency in the band in which their radios operate. Each radio operator also has a multi-band scanning receiver set up to receive the coordination frequencies *on each band*. For inter-organizational contact, one transmits on the coordination frequency on one band, and can hear the response on any of the coordination frequencies in any of the other bands. Various other communications nets (e.g., command nets, ground-to-air nets, hospital-to-scene nets, tactical nets) can also use shared frequencies with this method to effect inter-agency communication for various purposes.

This type of communication does have its limitations. It should be used only for emergency traffic and preceded by mutual communications agreements among the organizations using it. There also may be difficulty when a person's scanner is paused on another frequency when someone is trying to contact

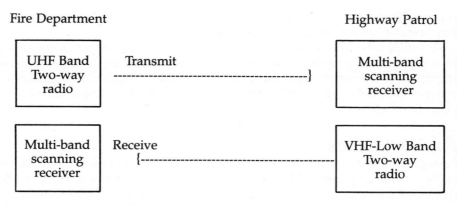

Figure 5-2.   Using scanners for "two-way" inter-agency communications.

him. Nevertheless, "two-way" use of scanners is one inexpensive means to carry out critical emergency communications among agencies whose radios do not operate on the same band.

Tone-activated Receivers

Specified frequencies can also be used to assure notification of various emergency or disaster-relevant agencies. In this case, each organization's dispatcher or radio operator has a tone-activated receiver (similar to a pager, or tone-activated weather frequency monitor). The area's communications center (or other pre-arranged organization) has a transmitter capable of emitting the tone alert, which activates the receivers. Different groups of organizations (e.g., all local fire departments, all local law enforcement agencies, all local hospitals), or all organizations having the receivers can be "toned," depending on the intended audience for the notification alert, update, or cancellation. One advantage of tone-activated receivers is that they are only activated when critical transmission of information is needed. They are, therefore, less likely to cause constant noise, resulting in people turning the volume down or off.

"Calling" Frequencies

In Southern California, fire agencies have designated a pre-arranged "calling-frequency." This frequency is used solely for initial radio contact and instructions. Subsequent to initial contact, other frequencies are assigned to incoming responders, depending on their function, assigned task, and operations area (ICS, 1980).

There are advantages to this technique that are applicable to inter-organizational disaster communications. One of the values of a pre-arranged regional or state-wide "calling" frequency is that it can even facilitate the initiation of coordination with *unexpected* or *unrequested* responders. Not only does a "calling" frequency allow resources to report their presence to those coordinating incident operations, but it can also be used to obtain essential initial information, such as:

— Are they, in fact, needed? (If not, they can be instructed to abort their response.)

— Where are they to report?

— How do they get to the check-in area?

— Are there any hazards that might be encountered en route?

— Are they to switch to another radio frequency for further communication?

One of the difficulties that has been experienced in a number of disasters is the initiation of communication between the incident site and responding helicopters. Large, bright cloth *ground-to-air signal panels* can be used for this purpose. Laid out on the ground at the disaster site, they are lettered with a message indicating the air-to-ground "calling" frequency, on which the aircraft can initiate contact with incident operations.

Appropriate employment of regional or state-wide calling frequencies requires the establishment of procedures for their use and pre-signed agreements prior to approval for operation on them. Also required is the availability of directories of the various calling frequencies.

Satellite Communication

The availability of sophisticated satellite capability can help alleviate communications problems, especially over wide geographic areas, in remote settings, or from the site to distant state or federal disaster assistance points (Kilijanek, 1981:67; Vines, 1986:10).

> **EXAMPLE:** *Tornado, Wichita Falls, Texas, April 10, 1979.* At the request of local officials, the State Division of Disaster Emergency Services contacted the Air Force for help with communications. Two communications jeeps were flown in which allowed a satellite communications link to be established between Wichita Falls and the state capital and simplified the state's ability to respond to local requests (Adams, 1981b:31).

This type of arrangement can be valuable in earthquakes too, since telephone transmission lines and microwave dish alignment are vulnerable to seismic forces (Seismic Safety Comm, 1979:7). Very small and lightweight portable satellite two-way radios make this type of communication an attractive tool (Cowley, 1985:223).

Common Mapping Systems

Multi-agency sharing of information about the present and predicted extent and location of disaster damage, secondary threats, vulnerable populations and structures, activities, transportation routes, and response activities generates the need for standardized maps and mapping grid systems. Yet, in some disasters as many as five or six different map types and map scales have been used by different agencies. This dissimilarity has caused considerable difficulty in communicating essential, geographically related information. The provision of standardized maps and familiarity with a standardized way (coordinate system) of describing locations on the maps are essential components to disaster communication.

Computers for Communication

Computers are not only useful for storing and analyzing disaster information, but also for sharing it. The graphic capabilities of modern personal computers have adapted themselves well to handling and transferring map data, the usefulness of which has already been described. The following are but a few examples of the uses of computers in disaster communications (Wohlwerth, 1987; Carroll, 1983; Carroll, 1985; Wallace, 1985; FEMA, 1987:27):

— Sharing and collating information about what agencies have responded and what resources they have dispatched.

— Locating and specifying procedures for obtaining "special" disaster resources (e.g., cranes, search dogs).

— Sharing information about the location, scope, and character of the disaster and damage that has resulted.

— Sharing information about the status of transportation routes, facilities (How many more victims can each shelter or hospital accept?), docking and landing sites.

— Generating and sharing predictions about weather and other expected conditions (areas subject to flooding, fire spread, movement of leaking chemicals, avalanche risk).

— Obtaining information on how to deal with a specific hazardous chemical.

— General "electronic mail."

SUMMARY

The communications equipment and procedures used by most emergency agencies are established primarily to deal with information flow *within* the organization. Disasters, on the other hand, pose heavy demands for *inter-agency* communication. To some extent, this can be facilitated by the availability of inter-agency radio networks, but some of the most difficult problems are "people" rather than "equipment" issues. The natural hesitancy to depend on and communicate with other organizations can be diminished in several ways. The critical information requirements of the various organizations involved in the disaster response need to be mutually understood, and the responsibility for gathering and disseminating it needs to be made clear.

PLANNING CHECKPOINTS

[   ] Does your disaster training program familiarize the members of various organizations with the tasks, methods, and responsibilities of other organizations likely to be involved in the response?

[   ] Do the organizations in your area have joint training sessions that deal with the common aspects of disaster response?

[   ] Have the emergency organizations in your area adopted standard terminology and procedures?

[   ] Has your area established inter-agency communications networks?

[   ] Does your area promote informal contacts among members of agencies likely to be involved in disaster responses?

[   ] Has your area developed a mutually acceptable division of responsibility for tasks likely to be necessary in a disaster?

[   ] Are your agencies' radio systems protected against disruption from common disaster agents? Are radio consoles and other components anchored to protect them from earthquake hazards? Are they protected from flooding? High winds? Is there a protected source of back-up power? An auxiliary antenna? Are protective and back-up precautions provided for the communications equipment of local commercial broadcast stations?

[   ] Is the responsibility for common disaster tasks predetermined on a mutually agreeable basis?

ADDITIONAL READING

Dynes RR, Quarantelli EL: Organizational communications and decision making in crises, Disaster Research Center Report Series No 17, 1977. Available from: Disaster Research Center, Newark, Del 19716.

Dynes, RR: Interorganizational relations in communities under stress. In: Quarantelli EL, Disasters: theory and research, 1978. Available from: Sage Publications, Ltd, 275 South Beverly Dr, Beverly Hills, Calif 90212.

Federal Communications Commission: Report on Future Public Safety Telecommunications Requirements, PR Docket No. 84-232: Phase 2, Private Radio Bureau, FCC, Washington, DC, 1985. Available from: International Transcription Services, 2100 M St, NW, Washington, DC 20036.

Joint Committee on Fire, Police, Emergency, and Disaster Services: California's Emergency Communications Crisis, California State Senate and Assembly, Sacramento, 1983. Obtained from the office of Senator William Campbell, California State Senate, Sacramento, Calif 95814.

Stallings RA: Communications in natural disasters, Disaster Research Center Report Series No. 10, 1971. Available from: Disaster Research Center, Newark, Del 19716.

CHAPTER 6

RESOURCE MANAGEMENT

Coordination of responding resources is a major problem in disasters. This photo shows emergency personnel and equipment outside the Kansas City Hyatt Hotel after the skywalk collapse of July 17, 1981. (Courtesy of the Kansas City Fire Department, Kansas City, Missouri.)

Disasters pose problems for resource management that are different from those in daily emergencies. Disaster tasks may require the use of resources (personnel, facilities, supplies, and equipment) from multiple organizations and jurisdictions and may also require the use of unusual resources. Much of the emphasis of disaster planning in many communities has traditionally been on the mobilization and reinforcement of resources. And, indeed, procedures for this purpose are important. More recently, however, it has been recognized that uncontrolled mobilization and over-response are common

103

problems in disasters. When they occur, coordination of response can be significantly complicated. This chapter discusses several problems that disasters pose for resource management and some approaches for handling them.

THE PROBLEM OF OVER-RESPONSE

One of the assumptions that often guides disaster planning is that the primary problem is one of mobilizing enough resources (Quarantelli, 1983:104). In fact, some persons have defined disasters as, "emergencies that exceed the available resources." (Orr, 1983:601; ACEP, 1976:901; ACEP:2–1,5–15; Cohen, 1982b:24; Comm on EMS, 1971; Holloway, 1978:60) However, this is not always the case. The problem of too many resources is coming to be recognized as a pattern which is found, at some time or another, in many disasters. When resources are present in greater amounts than needed, they can greatly complicate the already difficult problems of coordination and communication (Wenger,1986:v; Quarantelli, 1970a:384; Quarantelli, 1972:69; Quarantelli, 1983:78,80; Golec, 1977; Stout, 1981:42; Williams, 1956:660). In the more extreme cases, excessive influx of resources has even been observed to physically impede activity at the scene.

> **EXAMPLE:** Nine minutes after a tornado hit, an ambulance was dispatched to the scene. The EMT on board was asked to make an assessment and report back. After asking a couple of questions of people at the scene, the EMT estimated there were 150 injuries and radioed back, saying, "Send everything available; it's a big one." This assessment was made in 2 minutes, and ambulances arrived from all over the state. However, the estimate was totally wrong, and outside ambulances were not needed at all. Three times the ambulances necessary arrived, many of which got flat tires and blocked the roadways (Quarantelli, 1983:68).

> **EXAMPLE:** *Coliseum explosion, Indianapolis, Indiana, October 31, 1963.* "Literally hundreds of nurses, doctors, first aid volunteers, wreckers, cranes, station wagons . . . outside the fairgrounds made it difficult to get inside the Coliseum. A mass of humanity and equipment had converged and filled almost all the space within and just outside the building. . . . " Finally, the Chief of Police gave the order to bar any further traffic, including ambulances and first aid personnel, from entering the fairgrounds." (Drabek, 1968:19)

There are five main reasons why resource excesses may occur in disasters:

- The resources surviving in the disaster-stricken community are greater than expected.
- People react to disasters with a spirit of concern and generosity. Assuming that resources are deficient and the community is incapacitated, outsiders send resources into the disaster area in large amounts—even if they have not been specifically requested.
- The determination of responsibility and establishment of procedures for assessing and requesting the overall resources needed are often neglected.
- Because of the lack of clearly defined contact points, absence of compatible radio frequencies, non-functional or overloaded telephone circuits, and communications overload, it is often difficult for those offering help to contact someone who can tell them whether or not they are needed. Assuming it is almost certain that help is needed and that too many resources are better than too few, they choose in favor of responding.
- It is often difficult for the recipients of unsolicited assistance to refuse it.

Surviving Resources

After disaster impact, the basic resources still available in the community may be underestimated (Mileti, 1975:85; Faupel, 1985:58).

Manpower of Emergency Organizations

Police departments, fire departments, ambulance companies, and hospitals routinely operate 24 hours a day. By calling in off-duty personnel, these organizations may be able to actually double or triple their manpower. Furthermore, these organizations may be able to increase their capacity by shifting people from their regular tasks to more disaster-relevant duties. The police personnel assigned to records, crime prevention, and vice, for example, may be shifted to patrol or traffic control duties (Dynes, 1981:44,62; Quarantelli, 1972:69; Kennedy, 1970:354).

Food and Clothing

Historically, food shortages have not been a characteristic of disasters in the United States. Food supplies in households, retail groceries, and in wholesale warehouses has been sufficient to maintain a community for as long as several weeks. Paradoxically, disaster victims sometimes have eaten *better* than in normal times. Power failures thawed frozen food which then had to be eaten (Mileti, 1975:115).

> **EXAMPLE:** *Earthquake, Anchorage, Alaska, March 27, 1964.*
> After the Alaska earthquake of 1964, food was not a serious problem for anyone. This was in spite of the fact that some households lost a considerable amount of food from broken containers and thawed freezers. Only one family had to obtain meals from a kitchen set up in the neighborhood by the Army. Many wives pooled foods with their friends or relatives and cooked for the group on whatever stove was available. People needing food could go to the supermarkets where guards were patrolling or the clerks were cleaning up and ask for an item that was needed. If it could be found undamaged, it was freely given to the asker. Sometimes food came from unexpected sources within the community. One farmer who did not have enough feed for his chickens, killed 40, dressed and froze them. He gave several to needy families in one hard-hit neighborhood. Another family had about 35 dozen eggs which they were going to prepare for an Easter celebration by a local men's service club. Many of the eggs survived the quake and were given to the Salvation Army to distribute (Lantis,1984:24).

Medical Facilities, Supplies, and Personnel

Medical supplies are often available at nearby hospitals or wholesale warehouses (Quarantelli, 1972:69). Loss of hospital capability is not common in disasters. Historically, it is quite rare for American hospitals to be damaged or destroyed by the impact, or for them to be incapacitated by loss of water or power (Quarantelli, 1983:81). In a study of 29 major disasters, the Disaster Research Center found that supply shortages were experienced by only 6% of the hospitals, and personnel shortages occurred in only 2%. Many hospitals reported that they had more regular staff and volunteers than they could effectively use (Quarantelli, 1983:82,109). As stated by one researcher:

> "Unlike less-developed countries, the United States, except perhaps in very poor, rural communities, has enough skilled personnel and adequate medical facilities to respond to most disasters—short of truly catastrophic events. The main barrier that impedes effective EMS delivery in disasters is, rather, the insufficient level of awareness, education, and organization." (Tierney, 1985a:83)

As long ago as 1956, Raker observed that most hospitals taking care of disaster victims did not have to carry out a large number of surgical procedures (Raker,1956:35). This pattern can still be seen in recent domestic disasters of some magnitude.

> **EXAMPLE:** *Hyatt Hotel Skywalk Collapse, Kansas City, Missouri, July 17, 1981.* When two elevated walkways crowded with people collapsed and fell on patrons dancing below, 113 died, and 200 were injured (90 of whom were admitted to the hospital). Yet only 29 surgical procedures were carried out in the following 3 days, and the busiest hospital carried out only 6 of these during the evening of the disaster (Patterson, 1981:414; KC Health Dept, 1981:13; Orr, 1983:601).

There are two main reasons why the hospital operating room load is often not heavier:

— In general, most disaster casualties have minor injuries.

— Most disasters in the United States have not been very large.

In the disasters studied by the Disaster Research Center an average of only 10–15% of the casualties were serious enough to require even overnight admission to the hospital (Quarantelli, 1983:81; Golec, 1977:176). In most disasters, casualties pose more problems in their numbers than in their severity.

OBSERVATION

In a study of 29 major disasters, only 10–15% of the casualties were injured seriously enough to require overnight admission to the hospital; only 6% of the hospitals suffered supply shortages, and only 2% had personnel shortages.

Outside and Volunteered Assistance

When calamity strikes, people want to help. This desire to assist is manifested in a number of ways. Civilians in and near the disaster area become involved in search and rescue, giving first aid, providing food, shelter, and comfort. Sometimes the aid is given in spite of considerable risk to the provider. Those living farther away send food, clothing, medical supplies, and money. Surrounding governmental and public safety agencies send in personnel, ambulances, fire apparatus, helicopters, and other resources. These offers of assistance may come from distant locations in other parts of the country or even from foreign countries (Fritz, 1956:25; Bronson, 1959:102).

This flood of generosity may have some unforeseen consequences. Often this inundation of assistance is unsolicited and greatly exceeds the needs of the stricken community. When this happens, the inpouring of resources, however generously motivated, complicates the coordination of disaster response efforts.

In spite of the *perception* that disaster stricken communities are in need of resources, it may be very difficult for outsiders to get accurate information on *actual* needs. In part, this is due to the inherent uncertainty of the disaster situation. Additionally, it is because many communities have no generally accepted set of procedures for the multi-organizational collection and analysis of information about the disaster—so that resource needs can be accurately determined and resource requests coordinated. One factor that probably contributes to the quantity of *"unsolicited"* outside assistance is the difficulty the providers experience in determining who is in charge of the overall disaster response and who has accurate knowledge about resource needs.

OBSERVATION

The lack of a mechanism for outsiders to find out whether or not their assistance is needed may contribute to over-response in disasters.

Donations

One of the ways that generosity is manifested in disasters is in the flood of donations that frequently pours into the impacted area (Fritz, 1956:22; Williams, 1956:660; Moore, 1958:169).

> **EXAMPLE:** *Earthquake and Fire, San Francisco, April 18–19, 1906.* Outside aid began to enter the earthquake and fire-ravaged city before the ashes were cold. Food, medicines, cots, and blankets were the vanguard of hundreds of tons of relief supplies that would pour into the city. Within a few days, $8 million [$103,728,000 in 1983 dollars] had been raised to help San Francisco, and in the months to follow, another million would be added to that. Railroad cars arrived first from the coastal cities of Los Angeles, Seattle, Stockton, Vancouver, and San Diego. Within a week, trains arrived from all over. A steamer and a bargeload of provisions arrived from Sacramento along with the message:

"San Francisco can count on Sacramento for the last bit of bread and meat in the house, can draw on us for every dollar we have, and then you can have our blood if you need it."

Ogden, Utah, had no bread for days, because it all went to San Francisco. Boys of the Chemewa, Oregon, Indian School bought flour with their savings and baked 830 loaves of bread for San Francisco which was sent by Wells Fargo Express. The New York City Merchants Association ordered 14 freight cars of foodstuffs by phone to be sent from Omaha. In one month 1,800 freight-carloads of supplies came into the city. Even the entertainment world joined in to help. Barnum & Bailey contributed a day's receipts of $20,000 [approx. $260,000 in 1983 dollars]. Sarah Bernhardt held two benefit performances, and George M. Cohan sold papers on Wall Street, some of them going for as high as $1,000. In Los Angeles, men on the streets with megaphones begged for money for San Francisco. Within 3 days of the quake, Los Angeles had sent or was ready to send 75 freight cars of donations to the stricken city (Bronson, 1959:99).

In some cases, donations may pour into disaster-stricken communities far in excess of local needs, and the recipient communities may be unprepared to handle the deluge:

EXAMPLE: *Tornado, White County, Arkansas, March 21, 1952.* In 1952, a series of tornadoes struck six states. White County, Arkansas, was severely hit; forty-nine people were killed and 675 injured. The following day, large amounts of food and clothing donations began to flow into Searcy, the hardest hit area. A warehouse had to be found and opened to accommodate the donations that arrived in carloads, moving vans, railroad express trucks, planes and freight cars. A large American Legion auditorium was secured for storage, but it was filled to the ceiling within 2 hours. An auto parts building with a capacity of about 84,000 cubic feet was filled within the next 12 hours. Another building, covering half a city block, was filled by noon the next day. After that, tent warehouses were opened up, then a gymnasium and an additional warehouse. All of these storage areas were promptly and completely filled. The sorting and processing of this material took the efforts of over 500 volunteers working for 2 weeks (Fritz, 1956:23).

EXAMPLE: *Tornado, Waco, Texas, May 11, 1953.* The unexpected volume of clothing donations (an estimated 3½ boxcars full) created a problem because no provision had been made to receive it. A warehouse was opened, but the amount of arriving material almost crowded the workers out of the building. Clothing continued to arrive for a month, over 3 tons of it remaining in the Salvation Army warehouse after the disaster period had passed (Fritz, 1956:25; Moore, 1958:169).

Sometimes persons in one organization assume another organization is in need and make resource requests for it without confirming that the need exists.

EXAMPLE: *Tornado, Flint, Michigan, June 9, 1953.* Hurley Hospital (where 750 [approx. 80%] of the casualties were treated) had an adequate supply of blood on hand for the disaster. Nevertheless, a Red Cross volunteer, and independently, Flint's mayor called all the local radio stations to have an appeal made for blood donors. As a result, the hospital was suddenly deluged with 2,000 people eager to donate blood. This proved to be extremely disruptive to the hospital's disaster operations. At about the same time, the Red Cross was disrupted by a flood of donors bringing unneeded cots and bedding as a result of public appeals made by Hurley Hospital, the Mayor, and the Salvation Army (Rosow, 1977:167,169).

Large Numbers of Responding Organizations

Another factor contributing to resource-convergence is the large numbers of organizations that often respond (Mileti, 1975:121). There are several reasons why this may occur:

— Initially, the extent of the disaster is overestimated, resulting in requests from the scene to "send everything you've got."

— Local branches of national disaster relief agencies call in help from their regional headquarters.

— Local divisions of state and federal agencies have responsibilities in the disaster area.

— Federal and state agencies dispatch teams to study the disaster agent or the disaster response.

— When organizations outside of the disaster area hear of the event, they send reinforcements (even when not requested to do so).

Unsolicited aid may come from organized ambulance, rescue, and other emergency services (Quarantelli, 1983:71). Unsolicited aid is often offered to hospitals by off-duty medical and hospital personnel (Drabek, 1968:21). Physicians and nurses who are *not* members of the hospital staff also may show up and offer assistance (Quarantelli, 1970a:388).

In a number of disasters it has been observed that the response exceeded the manpower needs created by the disaster. In Quarantelli's study of 29 disasters, there was an average of only one patient transported per organization at the scene. (That is, one patient per *organization*, not per vehicle.) Furthermore, vehicles almost never made multiple runs (Quarantelli, 1983:22,71).

> **EXAMPLE:** *Airliner Crash, Kenner, Louisiana, July 9, 1982.* Unsolicited ambulance and rescue units came from as far as 70 miles away, but the disaster left only four seriously injured survivors. "More doctors and nurses than planned or expected arrived at the scene. Command personnel were unaware that they were even coming (and therefore could not cancel their response). While well intentioned, the flood of personnel and equipment overwhelmed site authorities . . . " making management and control difficult (Morris, 1982:65).

> **EXAMPLE:** *Earthquake, Coalinga, California, May 2, 1983.* "The initial emergency response by area public safety agencies and volunteers was commendable; however, the response came close to bordering on over-reaction. Within hours, the city had 30 ambulances and five medi-vac helicopters at their disposal, none of which were requested." Ambulances came over 100 miles from the San Francisco Bay Area without having received an official request by the Coalinga authorities (Seismic Safety Comm, 1983:81,86).

Volunteers

> "Whatever planning is undertaken, it can rarely prepare for the quantity and quality of volunteers that appear."
> —E. L. Quarantelli
> Disaster Research Center
> University of Delaware
> (Quarantelli, 1965:111)

Another cause of resources convergence is the large numbers of volunteers who often respond (Bush, 1981:1; Fritz, 1956:40). The United States has a long and vigorous tradition of volunteerism. Almost half of this country's ambulance services are volunteer (Dick,1982:67). Volunteerism is a tradition in the fire service, support of the Olympic Games, and numerous social and philanthropic causes. This altruism does not vanish in the face of disaster. If anything, it becomes stronger (Quarantelli, 1970b:625; Dynes, 1970a:436).

Not everyone anticipates the extent to which unofficial voluntary and un-solicited help is offered when a disaster strikes. In fact, within the stricken area, *more* potential volunteers become available, because schools and non-essential businesses close down (Quarantelli, 1972:69). In some disasters, more rescue and relief has been provided by unofficial civilian volunteers than by formal emergency or disaster agencies. In contrast to volunteers who participate in the response to routine emergencies, disaster volunteers are often unsolicited, and volunteer activity is unexpected (Raker, 1956:20; Williams, 1956:657; Barton, 1969:132,144,147,161; Dynes, 1974:30; Dynes, 1981:xviii; Quarantelli, 1983:36).

> **EXAMPLE:** *Tornado, White County, Arkansas, March 21, 1952.* After impact, 1,000 residents of a nearby town (representing 26% of the adult population) volunteered their services in the four medical centers during the first night (Fritz, 1956:43).

> **EXAMPLE:** *Tornado, Cheyenne, Wyoming, July 16, 1979.* During the first 2 hours after impact, an estimated 29% of the total search and rescue effort was completed by individuals who were not affiliated with any emergency organization (Drabek, 1981:119).

> **EXAMPLE:** *Floods, Texas Hill Country, August 1–4, 1978.* When three Texas counties suffered extensive flooding in 1978, two-thirds of those needing rescue or help received it from

Figure 6-1.   The Cheyenne, Wyoming tornado of July 16, 1979 is a good example of individuals not affiliated with any emergency organization helping out in an emergency. (Courtesy of Peter Willing.)

persons unaffiliated with any emergency organization (Drabek, 1981:68).

Difficulties with the volunteer response.
Organizations frequently have difficulty coordinating the efforts of volunteer workers with their own efforts. This is especially true when these people have never worked together before. Volunteers may have varying skill levels and lack familiarity with organizational routines or operating procedures. Organizations making use of volunteers cannot always count on the required task being completed, or if completed, it is uncertain with what efficiency, speed, or quality the task will be carried out. Furthermore, volunteers are not always familiar with the standard terms or routes used in communications. They don't know who to ask for what, or under what conditions (and to whom) to report difficulties (Quarantelli, 1970a:388; Quarantelli, 1983:21; Dynes, 1981:33; Killian, 1953:6,16; Faupel, 1985:52).

Benefits of the volunteer response.
In contrast to the picture painted above is the significant role volunteers play in decreasing the toll of death and destruction caused by disasters (Barton, 1969:132).

> **EXAMPLE:** *Tornado, Lake Pomona, Kansas, July 17, 1978.* On June 17, 1978, the showboat Whippoorwill, carrying 60 passengers and crew on Lake Pomona, Kansas, was struck by a tornado and capsized. At least a dozen nearby boaters rushed to the overturned vessel. One rescuer brought 15 to 18 people to shore who had been pulled from the water. He then returned to the Whippoorwill—and diving repeatedly into the water—located and helped to the surface several persons trapped below. Spontaneous action by civilian volunteers was responsible for all of the lives saved in this incident. Even if there had been an elaborate rescue system in the area, it is unlikely that it would have reached the victims sooner than did the volunteers (Drabek, 1981:53; Kilijanek, 1980:68).

> **EXAMPLE:** *Earthquake, Coalinga, California, May 2, 1983.* A report stated: "Local untrained citizens did most of initial search and rescue work, including control of utilities on a block-by-block basis." "The *immediate* community response of untrained citizen volunteer disaster service workers was vital to the fire suppression, search and rescue function and utility control to alleviate fire hazards. Without this responsiveness a much larger disaster would have resulted. . . . " (Seismic Safety Comm, 1983:97,98)

Figure 6-2.   After the showboat *Whippoorwill* was capsized by a tornado, June 17, 1978, the spontaneous search and rescue effort of nearby recreational boaters was responsible for all of the lives saved. (Courtesy *Topeka Capital-Journal*, Topeka, Kansas.)

INTER-ORGANIZATIONAL RESOURCE MANAGEMENT

It is important for response coordinators and planners to appreciate the tendency toward over-response in disasters. Although it is unlikely that this pattern can be completely controlled, effective measures can be taken to reduce it and channel it. In part, this can be accomplished by the development of procedures for the multi-organizational management of resources, so that needs can be accurately determined and requests coordinated. This helps to decrease exaggerated estimates of damages and injuries and the resultant appeals for mass assistance which may not be needed.

Overall Needs Assessment

A prerequisite to effective and efficient resource management is an accurate *overall* analysis of the disaster situation and the available resources. The data for this analysis must be collected from multiple organizations in order to get an idea of the "big picture." The failure to carry out this task has been a frequently observed problem in disasters (Parr, 1970:425; Mileti, 1975:80; Dynes, 1978:58; Quarantelli, 1983:65,114; Scanlon, 1985:123; Kilijanek, 1979:5; Rosow, 1977:136; Yutzy, 1969:118,156; Stallings, 1971:30). Often this is because it is unclear whose responsibility this is (Drabek, 1986:54). It is important to anticipate the fact that initial information about the disaster situation is often inaccurate (Dynes, 1974:77; Quarantelli, 1983:68). For this reason, *needs assessment has to be an ongoing procedure* that continues throughout the duration of the incident.

PRINCIPLE

Procedures for ongoing needs assessment are a prerequisite to efficient resource management in disasters.

Needs assessment involves two major processes: 1) situation analysis; and 2) resource analysis. Situation analysis is the collection of information about the extent and character of the disaster itself and the problems that have to be tackled. Resource analysis involves the collection of information about the resources needed to tackle the problems.

Overall Situation Analysis

Situation analysis difficulties in disasters.
The responsibility for overall situation analysis in disasters often is not clearly delineated. Even when situation analysis is carried out in disasters, it is usually done by individual organizations who seek out primarily that information of importance to their own organization's activities. Often, critical information possessed by one organization and needed by another is not shared (see Chapter 4).

> **EXAMPLE:** *Tornado, Flint-Beecher, Michigan, June 8, 1953.*
> Initially, the local post of the Michigan State Police got a report that the center of impact was at a drive-in theater. They were also under the impression that the direct road to the drive-in was blocked by debris and, therefore, impassable. Shortly thereafter, a fire truck from a nearby town worked its way past the drive-in, found that the road was in fact open and that there had not been a severe impact there after all. But what the firefighters knew was not communicated to the State Police. The State Police, assuming heavy casualties at the drive-in and that the direct route was blocked, sent badly needed ambulances there on a 2-hour drive on a roundabout detour. Furthermore, when the State Police discovered the correct situa-

tion, they did not inform the Red Cross, which sent a field radio unit to the drive-in (Barton, 1963:102; Rosow, 1977:136).

The prompt receipt of information about a disaster situation allows hospitals to start mustering and organizing their resources (Raker,1956:16,33). However, transmission of this information from the disaster site to community hospitals is a recurring problem. In 29 disasters, the Disaster Research Center found fewer than 12% of the cases where hospitals received useful information from the scene about the number of casualties to expect, or the type and severity of injuries (Quarantelli, 1983:67,91; Golec, 1977:174).

> **EXAMPLE:** *The Air Florida Crash, Washington, D.C., January 13, 1982.* At Washington Hospital Center's Medstar adult trauma unit, confusion reigned. Park Police officials instructed Medstar personnel that 4 or 5 victims would be arriving in two helicopters. When the patients failed to arrive, Medstar officials manned both radios and telephones in their effort to find out what was going on. "When an Army helicopter landed with a female patient suffering from hypothermia, a collapsed lung, multiple compound fractures, and internal burns from inhalation of jet fuel, she was taken to the hospital's 4th floor operating room—in order to reserve Medstar's single surgical bed for more critically injured victims. None ever arrived." (Goodwin, 1982:14)

> **EXAMPLE:** *Earthquake, Coalinga, California, May 2, 1983.* Hospitals in nearby Fresno received little information from the county emergency operations center. Apparently, the designated hospital radio notification (HEAR) system was not used (Seismic Safety Comm, 1983:86).

In a number of cases, the hospital's initial information was from the first arriving casualties or ambulances (Neff, 1977:186; Golec, 1977:173).

> **EXAMPLE:** *Metrorail Crash, Washington, D.C., January 13, 1982.* None of the major hospitals in the city were notified of the accident. The first notification that George Washington University Hospital received was from a paramedic who arrived with a victim of the Air Florida crash (which occurred 30 minutes before the Metrorail crash). The hospital never received any information regarding the number of casualties, type of injuries, or estimated arrival time. Most of what D.C. area hospitals learned of the disaster came from incoming EMS personnel or the news media. Furthermore, the hospitals were

not successful in reaching the appropriate officials to obtain additional information. This made it very difficult to determine the level of preparedness necessary for incoming victims (Edelstein, 1982:159).

Sources of information.   Disaster situation analysis may be accomplished by the use of multi-disciplinary damage assessment teams.

> **EXAMPLE:** *Earthquake, Anchorage, Alaska, March 27, 1964.* Damage assessment teams were organized at the Safety Building and dispatched to make a block-by-block reconnaissance. These teams were composed of 6 to 10 persons (e.g., a mechanic, an electrician, a plumber, a medical person). They were asked to assess damage, shut off utilities, search for survivors, and return with a report of their findings (Yutzy, 1969:120).

Alternatively, radio reports of situation data from each agency can be collected at a central point where it is collated and analyzed. The overall situation analysis is then composed into a report, including appropriate maps, which is made available to all responders.

Additional sources of information may include:

— Computer programs for forecasting fire spread, flood involvement, or spread of leaking hazardous materials.

— Weather predictions and satellite data from the National Weather Bureau.

— Air reconnaissance information including infrared detectors carried by aircraft to analyze fire hot spots.

— Computer data-bases with information on geographic features such as topography, access routes, helicopter landing sites, vulnerable structures, special features at risk (hospitals, nursing homes), and special hazards such as oil storage tanks, dams, and chemical storage sites.

Types of information needed for situation analysis.

Present conditions.   Important information includes that related to location and severity of damage; existing threats (fires, explosions, chemical spills, downed electrical wires, weakened structures in danger of collapse); numbers, locations, types, and severities of injuries; and numbers and locations of trapped victims.

Expected conditions.   Examples of factors that might influence expected conditions include: rate of fire spread; rate of river rise; seismic aftershocks; *tsunami* (tidal wave) or *seiche* (earthquake-generated wave in a lake or other closed body of water); hazardous spills due to earthquake; duration of the incident; weather influences; and downed power lines after storms.

Impact of expected conditions.   Examples of impacts of expected conditions include: evacuation areas; public shelter and feeding; need for sandbagging; possibility of further casualties; need for replacement personnel and reinforcements; need for feeding, sleeping, and sanitation facilities; need for fuel; and vehicle and equipment maintenance or replacement.

Overall Resource Analysis

Resource analysis difficulties in disasters.   Ascertaining what resources are present at the disaster is often quite difficult. This is because:

- Persons and organizations arrive at the scene without having been requested.

- Multiple organizations may independently request resources without informing the other organizations.

- It's not always clear whose responsibility it is to keep track of all resources involved in the disaster (Golec, 1977:169; Parr, 1970:426).

- It may not be clear to arriving resources who is responsible for overall disaster site coordination and to whom the resources should report.

Types of information needed for resource analysis.   Information needed for resource analysis includes data on what objectives need to be accomplished, what resources are needed to accomplish them, what resources are present and/or assigned, and what resources are available. When the situation analysis is complete, the results should identify those general problems that have to be tackled (incident objectives). These are broken down into specific tasks that are delegated to the various organizations present and their subdivisions. For each delegated task, the group responsible for accomplishing it must then indicate the resources it needs to do so. The indicated resource needs are then compared with resources present to assess the resources that need to be requested or reassigned.

Check-in areas.   One technique for obtaining and providing information on what resources are present is the use of check-in areas (Drabek, 1981:112).

EXAMPLE: *Tornado, Waco, Texas, May 11, 1953.* "As organization proceeded, three location points were set up at which personnel were to report. . . . These check points greatly facilitated the use of volunteers. When a particular type of skill was needed, it was often found that a person with that skill was actually present at one of the points. Furthermore, this prevented a situation that had occurred several times: a person would volunteer for a particular type of work and be told he was not needed; later, when he was needed, those in charge of personnel were unable to locate him." (Moore, 1958:16)

A useful arrangement is to have law enforcement agencies set up a security perimeter around the disaster site. At roadblocks, they can then divert incoming responders to a nearby check-in area located outside the perimeter. Such an off-incident reporting area has also been called a staging area or mobilization center (Brunacini, 1978; ICS,1983b; 1983). The person in charge of the check-in area has the responsibility for keeping track of those who arrive and conveying that information to the incident command post. As the responders are needed at the disaster site, they are requested from the check-in area by the command post. They are then assigned a communications frequency and told where and to whom to report as seen in Figure 6-3.

This type of arrangement has several advantages:

- It decreases radio traffic by allowing responders to check in and to receive a briefing and assignment in person.

- It allows a means to inventory and integrate into the system volunteers and unexpected responders.

- It keeps unneeded resources from congesting the disaster site.

- It allows responders to be integrated into the system even if they

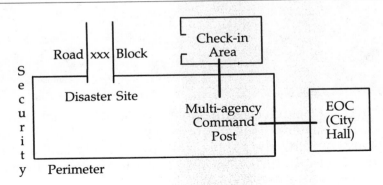

Figure 6-3.   The disaster check-in area.

have different radio communications frequencies, or if they do not know on arrival who is in charge or where to report.

- It prevents needed personnel from being excluded because they have not arrived with the usual "symbols of authority" which tend to grant passage through road blocks (red lights, official vehicles, uniforms, surgical scrub suits).

The technique, to be maximally effective, does require the prompt establishment of a security perimeter, and that those manning it know that the check-in area is functioning, where it is, and that incoming responders should be directed to it.

Figure 6-4.   Search dogs are an example of "special" resources needed in disasters. This photo shows Marcia Koenig and her dog, Bear, combing the rubble from the Wichita Fall, Texas, tornado of April 10, 1979. (Courtesy National Association for Search & Rescue. Photographer: Bob Koenig.)

Disaster resource-acquisition directory.   Disaster
tasks may require "special" resources—cranes, search dogs, satellite communications equipment, devices for listening for signs of life in the rubble, and
equipment or skilled personnel for cleaning up hazardous chemical spills.

> **EXAMPLE:** *Volcano Eruption, Mt. St. Helens, Washington, May
> 18, 1980.* A basic problem was establishing a communication
> network among multiple base camps, over 30 helicopters in the
> air, and the emergency operations centers in Vancouver and
> Olympia. The solution was accomplished by tapping a unique
> and recently developed resource. A C-130 aircraft carrying a
> specially designed communications jeep was requested

Figure 6-5.   "Special" disaster resources include heavy earthmoving equipment such as that in this photo of search and rescue
operations following the San Fernando Valley earthquake of February 9, 1971. (Courtesy Los Angeles County Fire Department, Los Angeles,
California.)

through the State Department of Emergency Services via the Air Force Rescue Coordination Center in Illinois and dispatched from March Air Force Base in California. The jeep was equipped to provide communications on most radio frequencies as well as by a NASA ATS-3 communications satellite. The C-130, which was also equipped with sophisticated communications gear that allowed it to monitor up to 65 aircraft at one time, was used as an airborne command post (Kilijanek, 1981:67).

Some of these "special" resources may not be part of the routine emergency inventory, and their access may not be covered by routine procedures (Drabek,

Figure 6-6.   Heavy mobile cranes are a "special" resource that may be needed in disasters. This illustration shows the search and rescue operations at the Hyatt Hotel skywalk collapse in Kansas City, Missouri on July 17, 1981. (Courtesy Kansas City Fire Department, Kansas City, Missouri.)

1981:xx; Drabek, 1986:185; Lantis, 1984:7; Gray, 1981:70; Ross, 1982:64; Scholl, 1984:287).

Locating "special" resources is facilitated by the existence of a current and complete inventory of available material and human disaster resources (Wenger,1986:16). The resource-acquisition directory is a list of such resources, where they can be located, and the proper procedures for requesting them. The directory may take the form of a written resource manual, index cards, or a computer data base. The latter has the advantage that it can be easily updated, and it can be shared by telephone line or radio interconnect. However, computers are also vulnerable to electrical surge or outage, and to earthquake shaking, and this should be taken into consideration. Examples of what might be included in a directory are illustrated in Appendix B.

Resource Allocation
Priorities and Sequential Interdependence

The allocation of disaster resources depends on the task priorities established for the incident. This may be affected by the fact that some tasks are "**sequentially interdependent.**" That is, the ability of one organization to complete its assigned task is dependent on another organization's accomplishment of a prior task (Drabek, 1986:178; Dynes, 1981:42).

For example, surgery cannot be initiated on a disaster victim unless that victim first reaches the hospital alive, and this depends on the skills applied by private ambulance paramedics in the field. But the paramedics cannot gain access to the victim until he is located in the rubble of a neighborhood of collapsed buildings. This requires the services of search dogs from private, volunteer organizations contacted through the sheriff's department. A crane to remove the entrapping rubble is then required, which the fire department obtains from a private construction firm listed in the local civil defense agency's disaster resource inventory.

Another example of sequential interdependence is the effect of disaster site activities on patient flow to hospitals. Typically, the distribution of casualties among the area hospitals is the result of decisions made at the scene, and hospitals are at the mercy of these decisions. The destination of casualties is often the result of independent decisions made by the various persons who come into contact with the victims, including police, firefighters, relatives, and civilian bystanders participating in search and rescue activities. Not rarely, this results in the bulk of casualties ending up at the closest hospital, while other facilities remain under-utilized (see Chapter 8).

Monitoring Task Progress and Resource Re-allocation

Allocation of resources based on the situation analysis can be complicated when, as so often occurs, the initial information obtained is inaccurate (Dynes, 1974:77; Quarantelli, 1983:68). Furthermore, attempts to accomplish disaster tasks are often plagued by unforeseen problems. Therefore, one cannot be certain that the various important disaster tasks will be completed as expected. This is particularly significant if the task is one upon which the accomplishment of a series of other crucial tasks is dependent. The appropriate countermeasure for this problem is the establishment of procedures to monitor the progress of the various tasks and to reassign resources to meet the conditions as they change.

Managing Volunteers

Management of disaster volunteers should take the following into consideration:

- Volunteers will respond—often in large numbers and on an un-solicited basis.

- While the presence of large numbers of volunteers makes coordination difficult, they play a significant—sometimes underestimated—role in saving lives and relieving the suffering that results from disasters. Volunteers may be able to provide needed services that are unavailable at the time from formal emergency organizations (Drabek, 1986:184).

- A large amount of the disaster aid from spontaneous volunteers occurs in the *early* period after impact when organized emergency and disaster agencies have not yet arrived in sufficient strength to provide the needed assistance. In many cases, these volunteers will turn over disaster work to formal, organized agencies when the latter can better supply the needed aid (Fritz, 1956:41).

- Difficulties with volunteers may be lessened if procedures are developed for integrating them into the formal organizational response.

For these reasons, effective coordination of disaster response must recognize and integrate volunteers, and someone must be vested with the responsibility for managing them.

One approach is to assign the responsibility for a check-in area where volunteers can report and an inventory made of their skills, abilities, and the equipment to which they have access. This may be at the same location as the check-in area for professional emergency responders or at a separate location. It is important that security personnel manning the road blocks and disaster site perimeter be aware of the check-in areas for volunteers and direct volunteers to these locations.

The effectiveness of volunteers can be enhanced by placing groups of them under the supervision of a trained member of a formal emergency organization (deputy sheriff, police officer, firefighter, National Guardsman) and assigning the group to carry out a specific task. The coordination of volunteer search and rescue efforts following disasters have been improved by this means.

> **EXAMPLE:** *Tornado, Waco, Texas, May 11, 1953.* Military personnel brought organization to the search and rescue efforts by incorporating civilian volunteers into their teams. These teams were composed of about 15 men under a leader and an assistant leader. In addition, there was one member with a walkie-talkie keeping track with the command post and with other teams (Moore, 1958:14).

> **EXAMPLE:** *Tornado, Wichita Falls, Texas, April 10, 1979.* By the time both the city emergency operations center and major field command post were in operation, many people in the area went to the command post to offer their assistance. The police captain in charge assigned members of emergency response organizations to direct search parties, each composed of 5 to 12 of these unofficial helpers (Adams, 1981b:24).

Members of emergency organizations who are assigned to lead teams of volunteers may find themselves acting as "instant teachers," explaining procedures as they go.

> **EXAMPLE:** *Earthquake, Coalinga, California, May 2, 1983.* "It was often necessary for me to stop and explain basic firemanship activities to the civilian on the fireground. When I asked for lines to be moved, I had to explain how to hold them, how to move them and how to put them together and operate them." (Seismic Safety Comm, 1983:116)

The coordination of volunteering organizations can be enhanced if each task assigned to volunteers is assigned to a group made up of members of a single existing organization. This preserves the existing coordination and communication procedures of intact groups and the advantages of working with familiar persons. Even if the organizations (for example, church groups, fraternal and sororal groups, clubs, unions, professional associations, private commercial firms) do not have specific emergency or disaster skills, their contribution to coordinated disaster activity is improved if their organizational structures are kept intact (Dynes, 1974:160).

In certain cases, volunteer organizations have special expertise applicable to the disaster situation. For example, organizations that provide special services to the disabled are well-adapted to handle the needs of the disabled in a disaster (Stevenson, 1981:45). The same might be said of organizations representing certain ethnic and language groups.

It is beneficial to have a process for registering volunteers as civil defense workers, so they will be covered by workers' compensation (Seismic Safety Comm, 1983:98).

Command Posts

A command post is a facility located *at the scene* of an emergency or disaster where management of site activities is carried out. In multi-organizational operations, coordination and resource management is facilitated when the leaders of the various responding organizations are located together in the same command post (Esch, 1982:167; Adams, 1980:13; Rosow, 1977:197; Moore, 1958:11). Frequently, however, different agencies involved in a disaster will each set up their own independent command posts (Best, 1980:22,25; Adams, 1982:54; Seismic Safety Comm, 1983:117). This is a situation that tends to isolate rather than unify response efforts.

Another problem at command posts is the absence of those with decision-making authority. Some organizational commanders neglect their overall command responsibilities and attempt to become involved in operations. Command refers to taking charge and making *general policy decisions* for the organization's overall response effort. Operations refers to the activities directly related to attacking the fire, rescuing the victims, giving emergency medical care, or arresting protesters. The effective commander does not allow himself to get bogged down in operations to the exclusion of other responsibilities, such as logistics (support) or planning. Rather, he delegates responsibility for the detailed management of these areas, while he provides *overall* direction (Brunacini, 1985:730,33,40; Wenger, 1986:31; 1983:16,67).

Emergency Operations Centers (EOCs)

In addition to command posts, disasters with multiple impact sites and large, complex disasters (e.g., those with significant state and federal involvement) often call for an emergency operations center (EOC). The EOC is usually established away from the disaster scene, often near governmental offices (e.g., city hall). In contrast to the command post, which is concerned with activities at the scene, the EOC establishes priorities for the distribution of resources among the various sites, and handles off-incident concerns (e.g., interaction with off-site facilities such as shelters; ordering of resources from distant jurisdictions or through state or federal disaster agencies) (FEMA, 1984b:i,A-1; Quarantelli, 1979a:23,35).

The idea of using an EOC to direct disaster response was initially a civil defense concept developed primarily with wartime use in mind. In more recent times, it has come to be used rather extensively in peacetime crises (Quarantelli, 1979a:9,11).

Based on the study of 180 local disasters, the Disaster Research Center concluded that in most cases when EOCs have been used, they have functioned fairly well (Quarantelli, 1979a:9; Wenger, 1986:iii). In fact, even when an EOC is not part of the disaster plan, one will often develop spontaneously anyway. The need for inter-organizational coordination and communication simply requires it (Drabek, 1986:186). Preplanning for an EOC, however, does seem to make it function a lot more smoothly, especially early in the disaster (Quarantelli, 1979a:15,16,18). Although EOCs generally work well, the following discussion will identify the areas where problems are most likely to develop if they occur.

The Number of EOCs

In some disasters, a number of separate EOCs appear, each involving a limited number of the participants in the total disaster response, and each dealing with a limited range of disaster problems. In such cases, there may be delays or deficiencies in needed information exchange among key officials located at different EOCs. In addition, persons having business with the EOC may be confused as to which one to contact. It is probably best in most cases to encourage the development of a single EOC, if maximal inter-organizational coordination is to result (Quarantelli, 1979a:12; Wenger, 1986:24,25,30). An exception to this rule might be the case where a disaster strikes in a number of counties within a state. Then it may be appropriate to have one EOC for each county interacting with a state-level EOC. A similar model might be used with several impacted states interacting with federal authorities.

Alternative EOC Sites

The Disaster Research Center found very few communities with plans for an alternative EOC location in case the original facility could not be used or had to be abandoned. Yet, the Center estimated that in as many as one fifth of the disasters, the necessity of moving the EOC became an issue. In three cases, this was due to flooding. The probability of this hazard could have been predicted by the examination of flood plain maps available from the Army Corps of Engineers or other sources (Quarantelli, 1979a:15).

Knowledge about EOC Location

In a number of cases studied, key persons and organizations were not aware that an EOC existed in the community and that it was part of the planned disaster response. Even if they knew that the plan called for an EOC to be activated, they could not indicate where it was supposed to be located. In some cases, this was because the EOC location was not clearly stated in disaster plans. In other cases, because it was not activated during disaster drills, officials were not familiar with it (Quarantelli, 1979a:14).

EOC Management Responsibility

Planning for an EOC needs to specify who is responsible for managing the facility. When this is neglected, questions arise regarding what space or communications equipment is to be allocated to what officials or organizations. Difficulties may be experienced when additional equipment or supplies are needed, or when decisions need to be made regarding who is allowed access to the facility—VIPs, non-governmental organizations, or the press, for example (Quarantelli, 1979a:22,25).

Presence of Those with Decision-making Authority

Coordination is hampered when representatives at an EOC lack the full authority, knowledge, and experience to make command and coordination decisions (Wenger, 1986:v,15; ICS, 1985a:62). Unfortunately, those assigned to the EOC often represent middle management levels of their respective organizations. This creates problems when urgent, high-level policy decisions are called for. The tendency of these managers is to follow the rules and regulations of their organizations rather strictly. On the other hand, flexibility, imagination, and initiative are needed to make the decisions required. Under these circumstances, top management representation is needed at the EOC. They have the authority and experience necessary to facilitate the type of innovative decision-making required (Quarantelli, 1979a:22).

Organizations Represented at an EOC

Even when most local key organizations are properly represented at an EOC, there are some that are often neglected. Private sector organizations like the Red Cross and private utility companies are among those that tend to be excluded.

The hospital is also one organization that is frequently in this category. Non-local organizations are not always represented at EOCs. Sometimes this is because they do not get involved early in the response, and by the time they do, the local EOC is already manned with most of the available space already taken. In some cases, local planning neglects coordination with non-local groups, and they are not notified that an EOC was being activated. In others these groups prefer to operate within their own familiar and routine channels of communication and authority (Quarantelli, 1979a:18,21; Wenger, 1986:30,31).

Attention needs to be given to anticipating the change in EOC representation during different phases of the disaster. For example, prior to disaster impact, organizations responsible for restoration and rehabilitation activities do not usually need to be present at the EOC. On the other hand, representatives of organizations involved in warning, protective, and preventive activities will be important. The scope of operations for many organizations will vary during the pre-impact, impact, and post-impact time periods. Accordingly, so will their need to be represented at the EOC (Quarantelli, 1979a:18).

The FIRESCOPE Multi-Agency Coordination System (MACS)

One system that holds promise as a model for multi-organizational emergency operation centers is the Multi-Agency Coordination System (MACS). FIRE-SCOPE (Firefighting Resources of Southern California Organized for Potential Emergencies) was funded by Congress after a series of devastating fires ravaged Southern California in 1970. Its mandate was to create a coordinated emergency response system for wildland fires. Two significant outcomes of this effort were the Incident Command System (see Chapter 7) and the Multi-Agency Coordination System. Whereas the Incident Command System was designed primarily for on-scene coordination, MACS carries out a number of region-wide functions (Chase, 1980; FEMA, 1987a:17). These include:

—information management,

—situation assessment, and

—resource allocation.

MACS is administered primarily through an off-incident emergency operations center called the Operations Coordination Center (OCC), which is the central information and resource coordination point for the system. The OCC maintains communications ties with numerous fire agency dispatch centers, local fire coordination centers, the National Weather Service, and, at times, the incident command post itself (Scherr, 1988). It also houses a set of computer data bases and programs to store, process, and exchange information (FEMA, 1987:27; ICS, 1987:1).

Information management.   MACS includes arrangements
for sharing inter-agency communications equipment and inter-agency coordination frequencies on a non-interfering basis during major, multi-agency incidents (Chase, 1980:11; ICS, 1980a; Scherr, 1988). Computer programs connected with local terminals provide for region-wide dissemination of up-to-date information (Chase, 1980:13; FEMA, 1987:17). Communications between the OCC, incident sites, and agency dispatch centers is carried out primarily via telephone line. As the situation requires, this is supplemented by various radio communications linkages (Chase, 1980:12,13; Scherr, 1988).

Situation assessment.   Several different types of information
necessary for situation assessment may be obtained by the OCC:

- Geographic information, including topography, fire risk, and man-made structures
- Current usage and availability of firefighting resources from the various agencies (Chase, 1980:4; FEMA, 1987:17)
- Current and predicted weather conditions (Chase, 1980:4; FEMA, 1987:17)
- Current and computer-predicted wildland fire involvement, including damages sustained; values of property threatened; and involved access, terrain, and vegetation (Chase, 1980:4,8)
- Current and predicted effectiveness of fire suppression activities (Chase, 1980:4; FEMA, 1987:17)

Resource allocation.   Local, state, and federal resources can be
requested for an incident and coordinated via the OCC. In major, multi-agency incidents, top command personnel from the participating agencies congregate at the OCC to coordinate operations (FEMA, 1987:18). Based on analysis of conditions at various incident sites, and the availability and location of resources, priorities are established for the allocation of resources. In addition, future resource needs for the incident(s) are anticipated and provisions made for their acquisition.

MACS administration and decision-making.   MACS
is designed so that it does not usurp the authority of any of the organizations using it. In fact, the participating (FIRESCOPE) agencies run the system. MACS does not *impose* decisions on the participants. Cooperation with MACS is voluntary, and policy decisions are by consensus. Decision-making is carried by a Board of Directors, Operations Team, Task Force, Specialist Groups, and

an Executive Coordinator. See Chapter 3 for more detail. (ICS, 1986; FEMA, 1987:17).

SUMMARY

In disasters, it is necessary to have established procedures for obtaining additional resources when they are needed. However, indiscriminate requests for resources can be detrimental. Many disasters are complicated by the over-response of resources, and this can greatly complicate the already difficult problems of coordination and communication. Procedures for pinpointing the specific types and numbers of resources needed are helpful in making the disaster response more manageable. Selected examples have been described in this chapter. The topic of the next chapter is the Incident Command System. One of the advantages of this system is the procedures it uses for inter-organizational resource management.

PLANNING CHECKPOINTS

[ ] Does your disaster plan and training program provide procedures for assessing the overall disaster situation and response needs? For disseminating this information among all the responding organizations?

[ ] Does your disaster plan and training program provide procedures for ascertaining all the resources at the disaster site? All the resources responding? All the needed resources available? The procedures for obtaining them? The time it will take for them to arrive?

[ ] Does your disaster plan and training program provide procedures for limiting the congestion caused by excessive responders at the scene?

[ ] Are there centralized procedures for requesting resources so that duplication of requests are avoided?

[ ] Are check-in areas a part of the disaster plan and training?

[ ] Whose responsibility is it to develop and maintain a disaster resource-acquisition directory? Does everyone know how to access this information?

[ ] Does your plan and training include procedures for incorporating and managing volunteers and unexpected responders?

[ ] Does your community have an emergency operations center (EOC)?

[ ] Are all the appropriate disaster organizations represented at the EOC?

[ ] Is the existence and function of the EOC well understood by all those who are expected to participate in it?

[ ] Do top level managers of disaster response organizations understand and accept the importance of their being present at the EOC?

[ ] How many EOCs will there be in a disaster?

[ ] Is the EOC vulnerable to disaster threats such as flooding?

[ ] Who has been designated as responsible for managing the EOC?

[ ] Are private sector organizations (Red Cross, Salvation Army, hospitals, private ambulance companies) represented at the EOC?

[ ] Who decides when the EOC is to be activated? What criteria are used?

[ ] How are all the EOC representatives notified that it is being activated?

[ ] Does representation at the EOC vary according to the phase of the disaster (e.g., warning phase vs. post-impact vs. recovery)?

[ ] Is there provision for the EOC to incorporate non-local or unexpected responders (e.g., the Environmental Protection Agency)?

ADDITIONAL READING

Drabek TE: The professional emergency manager: structures and strategies for success, 1987. Available from: Institute of Behavioral Science #6, Campus Box 482, University of Colorado, Boulder, Colo 80309.

Federal Emergency Management Agency: Emergency Operating Center (EOC). In: Job Aid Manual, SM-61.1, 1983. Available from: FEMA, PO Box 8181, Washington, DC 20024.

Federal Emergency Management Agency: Emergency Operating Centers Handbook, CPG 1–20, 1984. Available from: FEMA, PO Box 8181, Washington, DC 20024.

Federal Emergency Management Agency: Exemplary Practices in Emergency Management: The California FIRESCOPE Program, Monograph Series No. 1, 1987. Available from: National Emergency Training Center, Emergency Management Institute, PO Box 70742, Washington, DC 20023. Further information on FIRESCOPE can be obtained from: FIRESCOPE, Operations Coordination Center, PO Box 55157, Riverside, Calif 92517.

Federal Emergency Management Agency: Using the emergency operations center. In: Emergency program manager: An orientation to the position, SS-1, 1983. Available from: FEMA, PO Box 8181, Washington, DC.

Quarantelli, EL: Studies in disaster response and planning, Final Project Report 24, 1979. Available from: Disaster Research Center, Newark, Del 19716, (302) 451-6618.

State of California, Governor's Office of Emergency Services, Telecommunications Division: California On-Scene Emergency Coordination Radio Plan (CALCORD), 1986. Available from: O.E.S., Telecommunications Div., 2800 Meadowview Rd, Sacramento, Calif 95832.

Wenger D, Quarantelli EL, Dynes RR: Disaster Analysis: Emergency Management Offices and Arrangements, Final Project Report 34, 1986. Available from: Disaster Research Center, Newark, Del 19716, (302) 451-6618.

THE INCIDENT COMMAND SYSTEM (ICS)

by
Robert L. Irwin

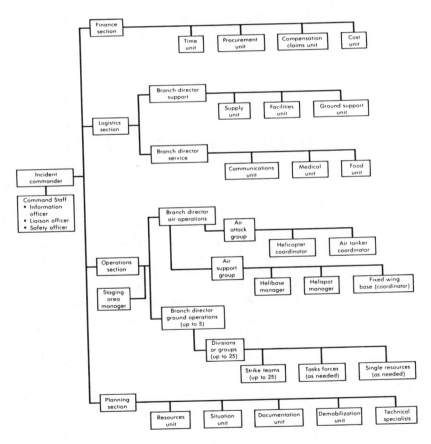

The incident command system (ICS). (Courtesy of Fire Protection Publications.)

The Incident Command System (ICS) discussed in this chapter was developed after a series of wildland fires caused death, damage, and destruction in southern California in 1970. Federal, state, and local fire services involved in the fire siege recognized hundreds of problems with their response and coordination during the fires. Most of the problems were quite similar to those described throughout this book. The fire services joined together in the FIRESCOPE Program to resolve those problems. The ICS was a major product of their joint effort. ICS is a management system, developed around specific design criteria and modern management concepts. There are five functions in the System, designed with a clarity that improves effectiveness, accountability, and communications. ICS uses an incident action planning process that is systematic and comprehensive; multiple agencies and emergency response disciplines can be integrated into a common organization using the process. The unified command concept used in ICS provides the most effective means of coordinating and directing multiple disciplines on major civilian emergencies.

DEFINING THE SYSTEM

The term "Incident Command System" has become popular across the United States in recent years. Hundreds of local jurisdictions and emergency response agencies have applied the term to a nearly equal number of organizational configurations. In many cases, the term has simply been applied to existing, traditional emergency procedures as a cosmetic approach to update or modernize an old way of doing things. Other terms, such as "incident management system" or "scene command" also abound. It is thus difficult to understand what they mean or to determine the exact configuration of any of these "systems" without some sort of definition. It is also dangerous to assume that one "incident command system" is as complete, or as effective, as any other.

In this chapter, the Incident Command System (ICS) under discussion is the version originally published by the FIRESCOPE Program (see Chapter 3) in 1982, and subsequently adopted by the National Wildfire Coordinating Group, the National Association of State Foresters, and the Federal Emergency Management Agency. A law enforcement version (LEICS) has been adopted and certified by the California Governor's Commission on Peace Officers Standards and Training (San Bernardino County, 1985).

The ICS described here is "a set of personnel, policies, procedures, facilities, and equipment, integrated into a common organizational structure designed to improve emergency response operations of all types and complexities."

ICS ORIGIN

In the fall of 1970, a series of devastating wildfires roared across southern California, burning over 600,000 acres and 772 structures in 13 days. Sixteen lives were lost during the period as a direct result of the fires. Thirteen of the largest fires were burning on federal, county, and city jurisdictions simultaneously.

California fire services had never faced such an immense challenge, and while many good works were accomplished during the disaster, it was clear to all involved that serious problems existed. Those problems were analyzed and documented in 1971 in an effort funded by Congress, led by the United States Forest Service, and supported by the state, county, and city fire departments that were involved.

The analysis identified hundreds of individual and specific examples of organizational weaknesses that were symptomatic of six major problem areas. It is informative to recognize that these problem areas were almost the same as those described throughout this book. They were:

- *Lack of a common organization.* More than 100 federal, state, and local agencies took part in the suppression efforts. There were at least a dozen different organizational structures in use, and these were frequently modified to meet contingencies. Terminologies (for position titles, equipment, facilities, and actions) were different for urban and wildland organizations, and at the local level even urban fire departments used different terms for the same items.

- *Poor on-scene and inter-agency communications.* Most of the 1970 radios were of single-frequency capability; scanners were rare; and federal, state, and local forces were operating in different frequency bands. On-scene supervisors could not contact subordinate units and frequently could not talk to those in command. Suppression units were essentially "on their own" and had to take independent actions that were not necessarily coordinated or effective. There were several cases where messages intended for units on one fire were received by units on a different fire, creating mass confusion. Agency dispatch centers could not communicate with each other, and major strategical events were not reported from agency to agency.

- *Inadequate joint planning.* Each involved agency did its own planning at is own chosen location. Forces were assigned on a unilateral basis, logistical support needs were ordered without knowledge of what other agencies already had available, and communications (as noted) were completely independent. On many of the fires there were separate and distinctly different strategic objectives created by different jurisdictions, at different times, and in different places. This resulted in diffusion of effort, safety problems, and failure to efficiently manage the incidents.

- *Lack of valid and timely intelligence.* None of the various organizational structures included elements charged with the specific duties of data and intelligence gathering. Information about incident character, size, and intensity were provided to decision-makers on a random and haphazard basis. No one had the defined responsibility to compile a comprehensive status report for any single incident, and obviously there was no means of analyzing the situations on the multiple fires. Much of the information upon which plans were made was 12- to 24-hours old, and forces were often dispatched to areas that needed no action. Other forces were not dispatched to critically important sectors.

- *Inadequate resource management.* As the combined consequence of all of the preceding weaknesses, resources were poorly managed. There were numerous examples of federal fire equipment heading one way "Code 3" (red lights and siren), passing state or local equipment going Code 3 the other way on the same road. Crews, engines, bulldozers, and other resources were lost, sometimes for days; no one knew where they were, and their potential effectiveness was lost. Some fires were overstaffed while others had no resources at all.

- *Limited prediction capability.* Since these fires occurred under extreme fire weather conditions and with the compounding challenges of both wildland and urban structural suppression, the expertise to predict future conditions (even 1 hour in advance) was lacking. No one knew where the fires were going, how many homes might be threatened, how many people should be evacuated, or where they might go if they were ordered to leave.

Examination of the analysis made it clear that there was need for inter-agency standardization and commonality, supported by modern technologies, if fire service performance was to improve. This led to the "design criteria" statement for a new system.

ICS DESIGN CRITERIA

The design criteria were developed before significant work began on developing the new system. This was done to assure that whatever the exact configuration of this new organization would be, it would be compatible with all of the requirements of a major emergency management system. The design criteria addressed a set of guidelines that included standard multi-agency organization, terminology, operating procedures, and communications integration. There were seven requirements placed on the design of the system:

- It must provide for effective operations at three levels of incident character: 1) single jurisdiction and/or single agency; 2) single jurisdiction with multiple agency support; and 3) multi-jurisdictional and/or multi-agency support.

- The organizational structure must be adaptable to a wide variety of emergencies (i.e., fire, flood, earthquake, rescue).

- It must be readily adaptable to new technologies that may become available to support emergency response and management.

- It must be able to expand from the organizational requirements of simple, daily incidents up to the needs of a major emergency.

- It must have basic common elements in organization, terminology, and procedures.

- Implementation of the system should have the least possible disruption to existing agency procedures.

- It must be simple enough to assure rapid proficiency of new users and to ensure low operational maintenance costs.

MANAGEMENT CONCEPTS AND SYSTEM CHARACTERISTICS

The fire services participating in the developmental effort (United States Forest Service; California Department of Forestry; California Office of Emergency Services; Los Angeles, Santa Barbara, and Ventura County Fire Departments; and Los Angeles City Fire Department) provided representatives who, collectively, had hundreds of years of emergency management experience. These people were practical, and familiar with all of the problems inherent in disaster response. They were all aware of "Murphy's Law" ("If anything can go wrong, it will go wrong."), and they wanted to keep Murphy away from the Incident Command System.

The fire services wanted to be sure the ICS was designed so that each agency would retain control over its own legal and fiscal responsibilities, agency roles, and organizational procedures. They wanted a system that would work well even with the participation of inherently different agencies and agencies from different levels of government (city, county, regional, state, and federal). Also desirable was a method for providing the best information management and maintaining order and effectiveness under crisis conditions. These needs led to yet another set of concepts and characteristics.

Concepts

Agency Autonomy

Throughout ICS, procedures are designed to protect agency (or jurisdictional) autonomy. The Unified Command concept was designed to encourage the close working relationship of diverse agencies while at the same time preventing "power plays" or "take overs" by larger or more assertive members. The system recognizes the legal and fiscal authorities of both primary and supporting organizations.

Management by Objectives (MBO)

The classic interpretation of MBO (Kast, 1974:171) is incorporated in the ICS planning process. The objectives set by Command must be "real" in the sense that subordinate positions agree that the objectives can be met. Command is required to adjust any objectives that subordinates state they cannot accomplish. This assures that plans are realistic and that Command is clearly aware of organizational limitations. It also increases the commitment of subordinate positions because those who help to design their own assignments have a greater motivation to reach objectives.

Unit Integrity

The organization is designed to keep people from the same agencies and emergency management disciplines together (i.e., police are not organizationally mixed with fire personnel; fire people are not assigned to public works). This concept improves the safety of the responders, makes it easier to keep accurate time records, and simplifies communication throughout the organization.

Functional Clarity

Each part of the organization is designed so that its members can concentrate on a primary assignment and not be unnecessarily distracted by other responsibilities. For example, the Operations Section does not have to be concerned with feeding, fixing flat tires, or obtaining special clothing. Units in the Logistics Section are activated to serve these and other needs so that Operations can put full energy into the basic assignment.

Characteristics

Effective Span-of-Control

Organizational supervisory positions are designed to provide supervisor-subordinate ratios that meet modern management practice. The general rule is five subordinate units per supervisory position, although allowance is made to vary this ratio under special circumstances. If tasks are relatively simple or routine, taking place in a small area, communications are good, and the incident character is reasonably stable, then one supervisor may oversee up to eight subordinate units. Conversely, if the tasks are demanding, taking place over a large area, and incident character is changing, then the span of control might be reduced to one supervisor per two or three subordinates. ICS is designed to provide the most efficient leadership possible under crisis conditions.

"Modular" Organization

The organization can be increased as an incident escalates in complexity, and it can be decreased as the incident comes under control. Following span-of-control guidance, an Incident Commander may respond initially with only a few units. As the incident grows, Command can add specific positions with specific assignments. Sections, Branches, Divisions, Groups, and Units (defined below) can be added. The complete (and rarely activated) organization will provide direction and control over 5,200 personnel. As the incident de-escalates, the organization can be reduced in a systematic manner, relieving those elements that are no longer needed. If appropriate, a demobilization unit can be staffed to assure prompt release of unneeded resources. Thus, ICS provides a means of adding and subtracting resources in the most cost-effective and leadership-efficient manner.

Common Terminology

There are several categories of "common" terminology:

Organizational positions.
Each position has a specific title (Incident Commander, Planning Section Chief, Branch Director, Division Supervisor). Although there are some necessary differences between the "fire" (ICS) and the "law enforcement" (LEICS) versions, the basic organizational structure is the same. For instance, the fire version has Section Chiefs while LEICS titles those positions Section Officer-in-Charge, or OIC. LEICS has "Armorers," "K-9 Units," and "SWAT Teams," while ICS has "Strike Teams," and "Air Attack" positions not in LEICS. The medical applications of ICS have introduced "Medical Supervisor," "Triage Leader," and "EMS Staging Manager." Each of these differences is justified by the requirements of the particular

discipline (fire, law, or medical). The differences, however, still follow a standard hierarchy (see Table 7-1).

Adherence to the hierarchical terminology, even though some special terms are needed, is what enables personnel from separate agencies or disciplines to understand and utilize ICS on multi-agency incidents.

Resource elements. Both ICS and LEICS define specific resources. ICS defines 16 "primary" resources (engines, bulldozers, helicopters) and 13 "support" or secondary resources (breathing apparatus, mobile mechanic, utility transport). LEICS defines 39 kinds of resources (SWAT team, light rescue team, coroner, patrol vehicle).

Defining the title and capability of specific resources, and having those definitions used throughout any particular discipline, has several advantages. First, resources can be ordered and managed to meet specific tasks; second, both the ordering and the sending parties know exactly what is needed; and third, the grouping of some resources into "teams" or "task forces" allows simplified resource accounting (see "Comprehensive Resource Management," below).

Defining resource elements and using those definitions throughout a jurisdiction or emergency response discipline is one important way to overcome the recurrent problem of incident managers ordering "everything you've got."

Facilities. Common terms are used to identify the facilities used at an incident, and each facility has a defined function. For instance, the Incident Command Post (ICP) is the location where Command functions are carried out.

Table 7-1.

Standard Organizational Positions	
LEVEL	TITLE
Command	Commander
Command Staff	Officer
Section	Chief/OIC
Branch	Director
Division/Group	Supervisor
Unit	Leader/Manager

The Incident Base is where personnel eat, sleep, and receive other care. The two facilities are not interchangeable in terms of function. Having common facility definitions and functions is another means of communicating and avoiding confusion; when personnel understand these functions and terms, they know where to go and what they will find at a given facility.

Integrated communications.

ICS/LEICS have a systematic process for making the best possible integration of available communications. Two standard forms, the "Radio Requirements Worksheet" and the "Radio Frequency Assignment Sheet" (see Appendix C) provide means to identify all available radio resources on the incident (mobiles, relays, base stations, and portables). These radio resources are then assigned to Command, Tactical, Support, Air-to-Air, and Air-to-Ground functions. These assignments abide by the unit integrity, agency autonomy, and functional clarity concepts of ICS, so no agency's radios are assigned to others without Command approval. The radio resources data are noted in Division Assignment Sheets and included in the Incident Action Plan (see Appendix C), so that all personnel on the incident have instructions on the available nets.

Comprehensive resource management.

ICS resource management procedures are designed to overcome the typical problems of too few, too many, lost, or mismanaged response forces. As with all other parts of the system, the resource management procedures are interrelated and compatible with the design criteria and management concepts.

Specific responsibility for resource status-keeping is assigned to the Resource Status Unit ("Restat") in the Planning Section. Restat is responsible for staffing "check-in" locations where all incoming resources fill out a check-in form (Appendix C). Data on resource status are continually updated, reported to Command, and used throughout the planning process.

Resources are managed either as single resources, task forces, or teams. The process simplifies status keeping and reduces span-of-control problems. Resources are monitored by three different status conditions: 1) "Assigned"—performing an active assignment; 2) "Available"—ready for immediate assignment; or 3) "Out-of-Service"—not ready for assignment. Status changes, major changes in location, and other data are recorded by a standard process that provides both Command and Planning with nearly real-time management information.

Two other extremely important components of ICS, the Unified Command concept and the Incident Action Planning Process, are discussed in detail later in this chapter.

OVERVIEW OF THE SYSTEM

There are 36 basic positions in the complete ICS organization (Fig. 7-1). The Command, Branch Director, Division Supervisor, Task Force Leader, Team Leader, and some other positions may be duplicated (following span-of-control guidelines) if necessary to expand the organization. With all positions filled,

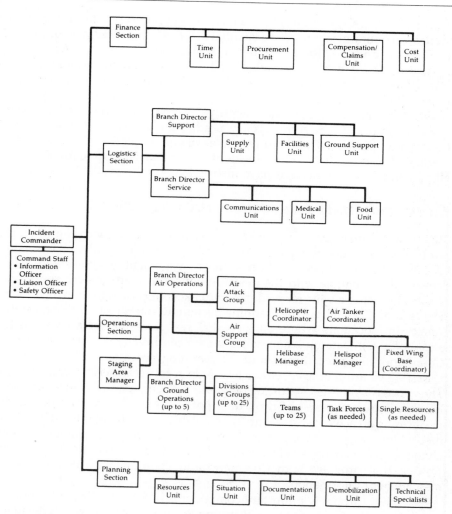

Figure 7-1.   Incident organization chart.

ICS can manage up to 5,200 people. It is rare that they all will be activated; only a major and very complex incident would require the full organization.

A basic rule of the organization is that the duties of any position **not** filled will be assumed by the next higher position. Thus, for moderately complex incidents where only perhaps one-third of the positions are activated, the complete range of duties and responsibilities would still be assigned to a specific person. For instance, if Command decides not to activate the Finance or Logistics sections, then Command must still be responsible for these functions. Or, if the Logistics Section Chief (or OIC) has only a moderate workload, a decision not to activate the Service and Support Branch Director positions may be made. In such case, the Logistics Section Chief/OIC would assume the duties of the positions not filled. This basic rule of delegation increases accountability and tends to encourage a stronger managerial perspective from Command and all Section Chiefs. The 36 positions are arranged to perform five functions: Command, Operations, Planning, Logistics, and Finance.

Command

Command responsibilities are **executive** in nature (see Fig. 7-2). They are designed to develop, direct, and maintain a viable organization and to keep that organization coordinated with other agencies, elected officials, and the public. Command responsibilities include:

—organizing to meet the needs of the incident,

—establishing incident control objectives,

—setting priorities for work accomplishment,

—assuring development of Command-approved Action Plans,

—approval of resource orders and releases,

—approval of public information outputs, and

—coordination with public officials and other agencies.

A key point about the command function is that the executive responsibilities cannot be ignored. Even though there may be only five or six responders on an incident and the Incident Commander may be quite involved in the actual "doing" work, the command function requires attention to organizing and managing.

The Incident Commander is supported by a Public Information Officer, Safety Officer, and a Liaison Officer as needed. These positions report directly

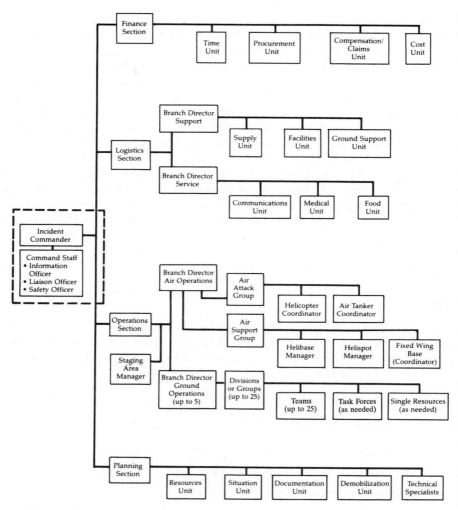

Figure 7-2.   Command section.

to Command and assist in fulfilling the duties of coordination with others and the overall safety of the organization's members.

Operations

The Operations Section responsibilities are of **line** nature (see Fig. 7-3). Operations is the "doer" in the organization, where the real work of incident control is

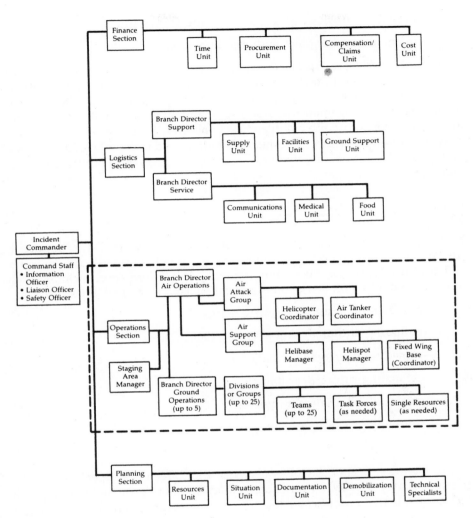

Figure 7-3.   Operations section.

accomplished. Operations is charged with carrying out Command direction. Responsibilities include:

—achieving command objectives,

—directing tactical operations,

—participating in the planning process,

—modifying Action Plans to meet contingencies,

—providing intelligence to Planning and Command, and

—maintaining discipline and accountability.

The most important observation that can be made about the disaster management failures documented in this book is that most disaster response organizations start and stop with the "doing" work. Earlier examples cite numerous instances where overall management has not been maintained, and only massive "doing" chores constitute the emergency actions. In ICS, the Operations Section activities—while certainly important—are integrated into a total managed system, and not a means unto themselves to the exclusion of all other chores that must be done.

Planning

Planning Section responsibilities are of **staff** nature (see Fig. 7-4). They are in support of Command and Operations, and designed to provide past, present, and future information about the incident. This information includes both resource and situation status on a real-time basis. Responsibilities include:

—maintaining accurate resource status,

—gathering and analyzing situation data,

—providing displays of situation status,

—estimating future probabilities,

—preparing alternative strategies,

—conducting planning meetings, and

—compiling and distributing approved Action Plans.

The Planning Section includes a position for "Technical Specialists." The position(s) may be filled by any qualified advisor(s) to provide Planning with technical data that are critical to incident management. In a flood situation, for instance, it may be necessary to consider public health and sanitation issues. A public health officer could be assigned as a Technical Specialist to provide professional advice. In the case of a building collapse, a construction engineer or the local building permit inspector might be used to advise Planning. The purpose of the position is to assure that plans are complete and realistic, regardless of the nature of the problem.

Logistics

Logistics Section responsibilities are also of **staff** nature (see Fig. 7-5). Logistics provides all of the personnel, equipment, and services required to manage the

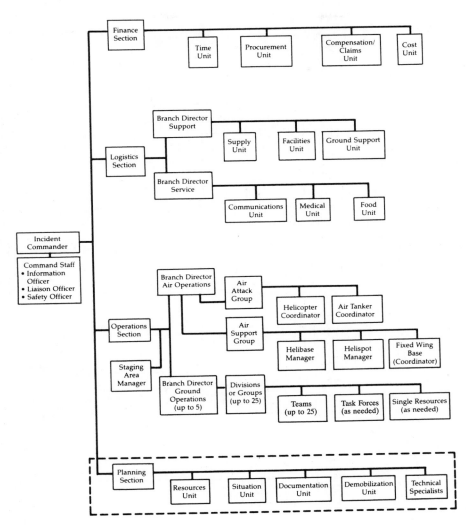

Figure 7-4.   Planning section.

incident. Following the "functional clarity" concept of ICS, Logistics is responsible for two subfunctions: Service and Support.

- *The Service Branch* is responsible for those tasks that "keep the organization going," such as communications, food services, and medical care for the incident (not public) personnel.

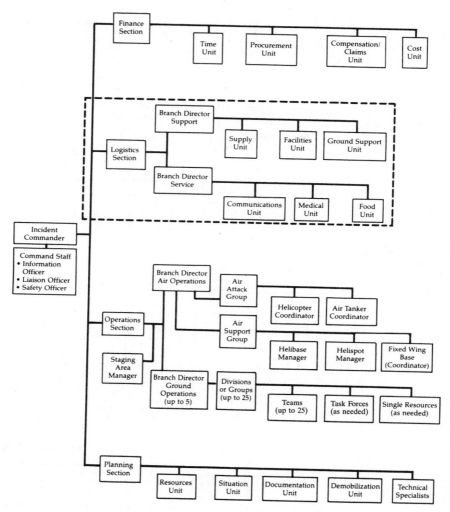

Figure 7-5.   Logistics section.

- *The Support Branch* assures that all parts of the organization can function; they provide adequate facilities, obtain supplies and resources, and service equipment.

It is important to note that once human, technical, and mechanical resources are obtained by Logistics, the management of those resources is turned over to Planning and Operations.

Finance

Finance is also a **staff** function (see Fig. 7-6). The Section is responsible for financial management and accountability on the incident. In keeping with the functional clarity concept, Finance authorizes expenditures in accordance with agency policies, but does not actually order or purchase anything. The Logistics Section obtains all needs after approval by Finance.

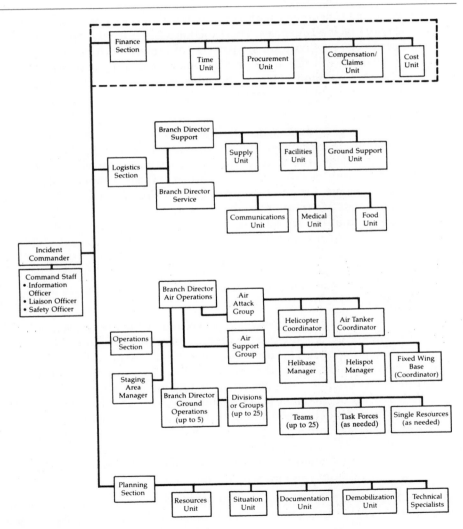

Figure 7-6.   Finance section.

Finance uses the Incident Action Planning process, the resource-status tracking, and the Logistics acquisition records to accomplish its accounting tasks. In addition to incident record keeping, the Section performs four other critical functions:

- *Disaster Relief Records* are used to coordinate with state and federal (FEMA) representatives and to assure that cost and damage records are prepared in proper format to assure reimbursement of private and public costs.

- *Contracting* is arranged with vendors for all services not available through involved agencies. This function assures legal preparation of contracts, sets cost rates, inspects equipment both before and after use, keeps use time on equipment and other contracted services (e.g., food caterers, portable toilets) and assures that services are delivered appropriately.

- *Agreements with other Agencies* are necessary during complex, multi-agency incidents when it is frequently quite cost-effective to share, or trade, resources. The classic example of this is in wildland fire suppression where one involved agency may have aircraft but lacks some kind of other resource, and one or more other agencies have specialized ground resources, but not aircraft. In these cases (and they happen frequently) the agencies' Finance Section Chiefs will agree that the aircraft use during the entire incident will be paid by the "owning" agency, and the specialized resources will also be used without regard to jurisdictional boundaries, and paid by the other agencies. Such agreements are also applicable to flood, hazardous materials, earthquake, and other types of incidents.

- *Injury and Damage Documentation* is prepared by the Compensation or Claims Unit, responsible for prompt recording of all injuries to incident personnel. This duty may be expanded to include civilian victims of the incident if the Incident Commander so directs. The unit is also charged with preliminary documentation and investigation of events that may lead to claims against any of the responding agencies. Such events might include damage to private property, personal injury, or any other kind of loss that could be construed to be a result of incident management activity. Documenting events when they occur, instead of weeks or months later, is a major task of the Finance Section.

In both ICS and LEICS, there are two checklists for the supervisory and subordinate positions of each of the five functional areas. There are general checklists showing the tasks all positions are accountable for on all incidents, and a specific checklist for detailing performance of each individual position. The checklists and other information about the system are included in pocket-sized "Field Operations Guides" (State of California, 1982) that can be provided to emergency response personnel as training tools and as reminders or references during actual incidents.

THE INCIDENT ACTION PLANNING PROCESS

Every emergency incident, no matter how small, requires some form of planning to control the problem. Better planning results in more effective and efficient response activities. ICS (and LEICS) use a planning process that meets the design criteria of "expansibility" from simple, daily activities up to the demands of a major emergency. It has been carefully designed to accomplish rapid, yet complete, planning for even the most complex of incidents.

For simple, routine incidents the process will be accomplished intuitively by the Incident Commander (the first arriving officer or supervisor). Even for the massive emergency where written Incident Action Plans should be prepared for every shift, the initial Incident Commander will probably start with an intuitive plan. However, ICS emphasizes that the mental and verbal procedures used in the early crisis should be rapidly replaced by the more formal and systematic planning process. Learning the formal process sets a mental pattern that allows for more complete application of the principles when intuitive planning is necessary.

Management-by-Objectives Framework

The Incident Action Planning process is derived from classic Management by Objectives (MBO) concepts below (Kast, 1974:171).

- Policy, objectives, and priorities are set by Command (the executive function).

- The organization required to meet the objectives is designed by Operations and Planning (the line and staff functions).

- Support and service needs, including communications requirements, are identified by Logistics (also a staff function).

- Financial abilities and constraints are considered. (This may be done by an activated Finance position, or reserved by Command.)

- A "reality-checking" review of the initial work is carried out. All participants in the process examine the tentative plan for completeness, feasibility, and capability to meet objectives. Results of the review are used to revise or strengthen the plan.

Forms Aid the Process

The experienced emergency responders who developed ICS spent over a year designing the forms that are used in the planning process. Their work was focused on preparing documents that would: 1) follow the MBO concept; 2) answer the questions, "What do we need to know?" and "What do we need to do?" on complex incidents; 3) be relatively easy to complete; and 4) be of real assistance, not just an exercise in paperwork, for incident personnel working under crisis conditions. All of those requirements were met.

There are two types or categories of forms used in the planning process. "Action" forms are those necessary to set objectives, assign the organization, and outline the tasks to be done. These are combined into the written Action Plan and provided to the personnel who will do the work. "Support and recording" forms are the remainder. They assist incident management by providing worksheets for systematic plan development, assuring that data and records are available and that resources are accounted for, integrating communications capabilities, and documenting decisions.

Many view the following list of forms and their applications as a formidable challenge, and "not quite worth the effort." That is not really the case. Trained incident managers can complete these forms in a very short time, even for complex incidents. The time required is materially shortened by the "fill in the blanks" nature of the forms and is materially offset many times over by the completeness of final planning and the effectiveness the process brings to emergency management. NOTE: All ICS forms are included in Appendix C. ICS and LEICS use the same form numbers throughout. There are some format differences between the two sets to accommodate the differences in disciplinary terminology. However, actual practice has shown that either ICS or LEICS forms may be used interchangeably because of their essential similarity.

Action Forms

Incident map (on form 201).   Page 1 of Form 201 is used for a sketch map of the incident if no better document is available. This is particularly valuable during the early stages of an incident to record situations, clarify thinking and communications about locations (for actions or problems), and to focus attention on overall objectives. This form can also be used to describe travel routes for resources (a "traffic plan") and locations of special facilities such as casualty collection points or evacuation centers. More sophisticated maps should be used for detailed planning if they are available. Pages 2, 3, and 4

of Form 201 are used to provide documentation on simple incidents and as a briefing format for succeeding Incident Commanders and other overhead personnel if the incident escalates.

Incident objectives (form 202).

Form 202 is the key to effective action. It is the initiator of the planning and control process and the place where Command begins to form and direct the organization. The form allows Command to describe all desired objectives and priorities.

Organization assignment list (form 203 or 207).

Form 203 (or Form 207) shows who has been assigned on the incident. It shows who's in charge and details reporting relationships. It also serves as a sequential record of the resources available by time period.

Division assignment list (form 204).

Form 204 provides detailed instructions for incident personnel. Information on the form specifies resources assigned, their configuration, and who does what. It is the place where actual tasks necessary to meet Command objectives are described, and may be used to further define priorities. Completed forms assist the "reality checking" phase of MBO by making span-of-control and communications decisions visible. They also assist in this regard by forcing another examination of available capabilities compared to objectives.

Completed forms are distributed as part of the Action Plan. All Sections, all Branch Directors, and all Division Supervisors have forms showing the resources under their direction, and the tasks assigned to those resources.

Communications plan (form 205).

Form 205 is one of the major tools that can bring order out of chaos on complex incidents. Its preparation and use improves multi-agency communications regardless of the types or capabilities of the involved radio systems. Preparation of the 205 is facilitated by completion of Form 216, described below.

Medical plan (form 206).

Form 206 is primarily intended to serve incident personnel. However, on incidents where medical assistance to the public is required, the form can serve "double duty" as an attachment to Medical Division Assignment Sheets. Conversely, in the case of a major multi-casualty incident, one of the Medical Divisions could be assigned the additional duty of caring for incident personnel, using the information from a Form 206 prepared for that purpose.

Organizational chart (form 207).

Form 207 provides a more visually detailed picture of the organization. It can be used in place of Form 203 (the organizational assignment list).

Support and Recording Forms

Incident status summary (form 209).
Form 209 provides a summary of current status. The form serves Command as an overview of the incident and may be used to forward details to local, state, and federal agencies interested in incident details and control progress. It may also be used (along with the entire Action Plan) as a briefing document for the media and elected officials.

Check-in list (form 211).
Form 211 is a basic tool for the Planning, Finance, and Logistics Sections. It provides data on all authorized resources on the incident and can be used very effectively to weed out those forces or persons who have simply gravitated to the incident because of its magnitude or notoriety. Item 5 on the form ("Order/Request Number") serves as an indicator of legitimacy: if the resource has been requested by Command there will be some kind of record of that request; if the resource is a voluntary response, this form will define it as such.

Unit log (form 214).
Form 214 is prepared by all assigned Units, Division Supervisors, and Branch Directors. It provides a record of actions, problems, and intelligence for future planning and a record of past events. It also assists in maintaining accountability.

Operational planning worksheet (form 215).
Form 215 is a valuable tool for Action Plan preparation and overall management response to any incident. Command objectives are listed, and the resources "required," "have," and "need-to-order" are shown. From this worksheet, and the process of its preparation, Command, Planning, Operations, and Logistics gain valuable management information. The reality of objectives (shown in the "Work Assignments" column) may be checked against resource availability, the total workload estimated, assignments further clarified, and the resource deficits, if any, recognized and corrected, if possible.

Radio requirements and frequency assignment worksheets (forms 216 and 217).
Forms 216 and 217 are the initiators of Form 205 (The Communications Plan). Block 5 of Form 217 ("Radio Data") may be modified to show the radio availability from any group of agency disciplines. Any qualified communications technician will be able to prepare this form quickly, given a general familiarity with agencies involved in even the most complex incidents. This information is then adapted into form 205 by the Logistics Section for use in the Action Plan.

Support vehicle inventory (form 218). Form 218 is prepared by the Logistics Section to provide records and maintain availability information on support and service vehicles. It is a tool for Finance and serves Command, Planning, and Operations by showing the authorized vehicles on the incident.

Air operations summary (form 220). Form 220 records air operations details. The Operations Section uses this form to manage aircraft in a manner that provides the best possible coordination between air and ground forces. Finance also uses the form in cost accounting.

UNIFIED COMMAND

Why Unify Command?

More than 90% of emergencies that occur daily in the United States are readily managed by local agencies using only their own resources. On a small percentage of emergencies, the responsible agency may exhaust its own resources and call on neighboring jurisdictions for assistance. Many agencies are experienced with these "automatic aid" responses and assist each other on a routine and problem-free basis. These incidents do not call for Unified Command and are best handled under a single command structure.

However, about 5% of all emergencies become serious enough to require the response of several agencies, each with its own legal obligation to perform some type of action, not just assist their neighbor. It is in these critical, multiple-involvement emergencies that Unified Command is called for. Some examples:

- *Incidents that affect more than one geographical jurisdiction.* The classic example is of a wildland fire starting in one jurisdiction and burning into one or more others. Floods and hazardous-materials incidents could be similar. The incident is essentially the same challenge in each jurisdiction, but the political and geographic boundaries mandate multi-agency involvement.

- *Incidents that affect more than one functional jurisdiction.* Major commercial airplane crashes are an example. The crash occurs in one geographical jurisdiction, but will involve fire suppression, law enforcement, medical response, Federal Aviation Administration, National Transportation Safety Board, and perhaps other agency response. All of these entities have different missions to perform, all at the same time, and all in the same place. The different functional roles, or statutory obligations, bring about multiple involvement.

• *Incidents affecting geographical and functional jurisdictions.* These are typified by the Mt. St. Helens volcanic eruption and the Three Mile Island nuclear accident. In these types of incidents, large numbers of federal, state, and local agencies become involved. The emergencies cross geographical boundaries and overlay multiple functional authorities.

In today's world, the public, private, and political values at risk in major emergencies demand the most efficient methods of response and management. Meeting this demand when multiple and diverse agencies are involved becomes a very difficult task. The Unified Command concept of ICS offers a process that all participating agencies can use to improve overall management, whether their jurisdiction is of geographical or functional nature (Irwin,1980).

What is Unified Command?

Unified Command is the first consistent, systematic means of organizing a variety of autonomous civilian agencies into one concerted emergency response effort. The concept offers uniform procedures that enable all involved agencies to perform their roles effectively. Unified Command overcomes many inefficiencies and duplications of effort that occur when functional and geographic jurisdictions, or agencies from different governmental levels, have to work together without a common system.

Unified Command is deeply rooted in ICS concepts and characteristics. It follows the same MBO planning processes, respects agency autonomy, maintains functional clarity, and provides a common management framework for action.

The goals of the Unified Command concept are to:

—Improve the information flow between the agencies involved.

—Develop a single collective approach to the management of the incident.

—Reduce or eliminate functional and geographical complexities.

—Optimize the efforts of all agencies.

—Reduce or eliminate duplications of effort.

These are practical goals. They have been achieved with relative ease on actual incidents involving multiple fire agencies, incidents requiring fire and law enforcement coordination, and emergencies that included fire, law, and medical disciplines. As the ICS becomes more completely implemented by agencies across the country, the goals will be met with greater regularity and greater effectiveness. When that happens, many of the consistent disaster management failures documented in this book will begin to disappear.

ICS Characteristics Pertinent to Unified Command

The Incident Command System is based on commonality. The commonality is a major departure from the traditional ways agencies have operated, and it creates significant opportunities for improvement over old methods. When agencies involved in a major emergency use ICS (the same organizational structure, the same terminology, and the same management procedures), there are few, if any, differences in operations. In essence, they are "one" organization, and can be managed as such. Instead of several command posts operating independently, the total operation can be directed from only one location. Instead of preparing several sets of plans (with no guarantee of coordination among them), only one set need be prepared to inform all participants. In place of several logistical and communications processes, only one system of collective and integrated procedures is used.

These five ICS characteristics (one organizational structure, one Incident Command Post, one planning process, one logistics center, and one communications framework) create a strong synergy. By meeting and working together at one location, preparing a single plan of action, and using other common procedures, the senior officers (Unified Commanders) from many agencies bring their collective powers to bear on the incident. They are able to share information, coordinate actions, improve resource utilization, greatly improve communications, and rapidly cope with changing incident conditions. This unified effort is supported and reinforced by the ICS Planning Process.

The Planning Process for Unified Command

The planning process for Unified Command is the same as for single Command, except that more people are involved. The process follows the MBO sequence, uses the same worksheets and forms, and allows for both functional and geographic response authorities to combine objectives and actions.

The process starts with documentation of each Commanders' objectives just as though it were a single-agency incident. These objectives may be widely different depending on incident character, agency roles, and other factors. It is extremely important to understand that these separate, and perhaps diverse, objectives do not have to be forced into a consensus package. Unified planning is not a "committee" process that must somehow resolve all differences in agency objectives before any action can take place. It is, however, a "team" process, and that promotes open sharing of objectives and priorities. Through the process, the team formulates collective (which is significantly different than "common") directions to address the needs of the entire incident.

Once collective objectives and priorities are documented, the process continues as it would for single-agency involvement, except that all agencies are included:

- The organization is designed to utilize multi-agency resources according to all span-of-control, unit integrity, and functional clarity guidelines.

- Support, services, and communications requirements are obtained and assigned.

- Branch, division, and unit assignments are detailed.

- Financial considerations are defined and agreements are documented.

- "Reality checking" is accomplished by staff of all agencies.

The developed multi-agency plan is returned to the Unified Commanders for approval. Again, it is important to understand that the individual Commanders in the group only approve those portions of the plan that affect their agencies.

Unified Command Configuration

In addition to all of its other attributes, ICS is a common-sense system. It is designed with a great deal of inherent flexibility. This allows modification of the on-scene organization to meet specific conditions, complexities, and workloads for different incidents. There are also various ways that a Unified Command group may be formed. The guidelines for deciding who should be in command are simple and apply at any level of incident complexity:

Agency Role

Responding agencies will be filling one of two roles. They will be either jurisdictional, with direct statutory responsibility and authority, or they will be supporting agencies who have been called for help.

Only jurisdictional agencies with statutory responsibility on some part of the incident can assign one of the Unified Commanders.

Agency Authority

The agencies who assign Commanders must have the authority to order, transport, and maintain the resources necessary to meet Command objectives. This authority is not dependent on size or budget level since even very small agencies may participate in a Unified Command. It is dependent upon legitimate capability to pay the bills. (In the case of small agencies, this capability may come from state and federal assistance, but is nevertheless the required capability.) Only agencies with fiscal authority may assign one of the Unified Commanders.

Applicability

These guidelines apply equally to multi-geographical, multi-functional, and multi-geographical-functional incidents. The guidelines can and should be modified to meet exceptional conditions. An incident of disaster proportions will involve state and/or federal agencies, and officials from those government levels may be appropriate members of the Unified Command Group.

Alternatives to Command Participation

There is a practical limitation on Unified Command participation. Once a group exceeds about eight persons, the effectiveness of that group begins to deteriorate. ICS concepts recognize this and recommend that no more than eight people fill the Unifed Command Group. During incidents where more than eight agencies have legitimate legal and fiscal authority, there are alternative ways to encourage total participation without having all in command. These and other alternatives have been used successfully on multi-agency and multi-disciplinary incidents. It requires training and experience to make the process work effectively. Pre-incident meetings, planning, and agreements facilitate the process. Two of the most popular alternatives to participation in the Unified Command Group are:

Deputy Incident Commanders

Agencies with limited involvement may choose to fill their commitment to the incident with a Deputy, rather than a "full" Commander. This will enable adequate input from that agency into the planning process, protect the agency's autonomy, and provide significant support to the unified effort.

Subordinate Positions

For smaller jurisdictions involved in a major emergency, it may be appropriate to designate that agency's area or function as a Branch, Division, or Group, and place a senior officer of the agency in charge. The officer (now a Director or Supervisor in the organization) will be an integral part of the unified effort and take part in the planning process. At the same time, he is fulfilling "at home" responsibilities, probably with his own forces, and serving his jurisdiction.

MANAGING MEDICAL RESOURCES

The function of the Medical Unit (see Fig. 7-5) is frequently misunderstood by persons not familiar with ICS. Medical professionals, in particular, express concern that such an important function seems to be placed in a subordinate role. It is important to understand that these concerns are unfounded. The Medical Unit's role is to take care of incident personnel, only. Very early in ICS development this was called the First Aid unit, but the title and the functions required were changed quickly to assure that incident personnel with more serious injuries could and would have adequate medical care. The intent and purpose of the Unit is to provide medical attention to responders that are part of the incident organization.

If an incident involves casualties that are victims of the emergency itself, then various forms of a medical response organization can be assigned. Medical entities will fit in any (or all) parts of the system, depending on the character of the incident. A public health officer or other M.D. could be the Incident Commander under some circumstances, or might be a member of a Unified Command Group. On major multi-casualty events, one medical representative could be the Operations Section Chief, others could be Branch Directors or Division Supervisors. Still other representatives could be in the Planning, Logistics, and Finance Sections. Groups of ambulance and paramedic personnel can be designated as Teams or Task Forces for just about any incident involving injuries.

At any level of severity the ICS concepts of modular development, functional clarity, and unit integrity will hold true for medical applications, as they do for other types of incidents. The organization can be increased to meet the needs of the event (see also Chapter 8, and Fig. 8-1). Some examples include:

- *Medical Emergency* (5 to 15 casualties): This level of severity could be managed by assigning a Medical Division Supervisor with a Triage Team, Treatment Team, and Transportation Team.

- *Major Medical Emergency* (16 to 50 casualties): In this case, the organization would be expanded by adding Officer and Unit Leader positions to assure that all required functions receive appropriate attention. Note that a Medical Communications Leader position is assigned to coordinate scene-to-hospital(s) communications.

- *Medical Disaster* (more than 50 casualties): The organization is expanded further by adding additional Medical Divisions. A Medical Branch could be added if more than two Divisions were required. If there were need for more than five Divisions, a second (or third) Medical Branch could be established to further increase the response and still stay within the span-of-control guides. In such cases the Medical Communications Leader position could be assigned to the Branch Director, and reduce the number of those positions at the Division level.

Medical applications of ICS can bring increased effectiveness to the discipline. As the ICS becomes more established with fire and law enforcement agencies across the nation, the medical discipline will find more opportunities to adopt the system.

INTEGRATING VOLUNTEER EFFORTS

It has been well established that volunteer efforts can both help and hinder emergency response agencies. The help comes in the form of immediate energies and work accomplishment. The hindrance comes from unmanageable (or unknown) numbers of volunteers, poorly directed work, and a general lack of control. All of the helping aspects of volunteer involvement can be accentuated, and all of the hindering dynamics can be reduced or eliminated by appropriate use of ICS.

For example, the modular flexibility of ICS can incorporate volunteer Units, Teams, Task Forces, and perhaps even Divisions. A qualified agency Division Supervisor can easily manage up to 30 individual volunteers, or up to a 100 if they are arranged in 20-person crews. A Branch Director could oversee the effective work of about 500 people under good conditions. The possibilities for integrating volunteers is essentially unlimited, provided the agency supervision is available. A few of those possibilities are search and rescue, sandbagging, evacuation alerting, road construction, and firefighting. The key element is supervision and fitting the resources into the organization. That requires Command attention to managing the organization, and brings us back almost full-circle to the responsibilities of the five functions in ICS.

In cases where volunteer efforts need to be managed, Command must recognize the situation and set reasonable objectives for those efforts. Command and Planning must develop the organization to provide supervision and clear direction to the volunteers. Planning must also inventory the volunteer resources through a retroactive check-in procedure and include them in the Incident Action Plans. Logistics must be able to service and support the resources and set up communications through existing agency, or perhaps "ham" (radio amateur) capabilities. The Finance Section should assure that volunteers are physically capable of doing the assigned tasks, are paid if so directed, and are properly compensated for any incident-related disabilities.

If volunteers are managed in this way, then the public agencies' response efforts will be more effective. If volunteers are not managed, then the typical problems and inefficiencies associated with their involvement will continue.

SUMMARY

After the 1970 fires, southern California fire services recognized that their experience included the same theme of weaknesses that are described after most disasters. They recognized that those weaknesses could be corrected if a systematic process for managing multiple and diverse resources were developed. The fire services described criteria and adopted modern management concepts that would reduce or eliminate the problems. The resulting system, ICS, was designed to cope with the basic causes of disaster problems.

ICS provides ways to quickly perform situation analysis and to use the analysis as a basis for realistic planning and actions. The organization integrates multiple resources into definitive functional efforts. It provides for direction and management of multiple disciplines and different government levels under crisis conditions; it improves communications; and it increases the effectiveness of all involved. The planning process brings order out of chaos, and the step-by-step use of helpful forms makes the process systematic and thorough. Unified Command procedures protect agency autonomy.

Major law enforcement and medical agencies in various parts of the nation are adopting the system without changing its basic configuration. This testifies to the fact that ICS is no longer viewed as a "fire" system and is now seen as it was intended to be—a management system.

ADDITIONAL READING

Exemplary Practices in Emergency Management: The California FIRESCOPE Program, Monograph Series No. 1, FEMA 117, 1987. Available from: Federal Emergency Management Agency, National Emergency Training Center, Emergency Management Institute, PO Box 70742, Washington, DC 20023. Free.

FIRESCOPE Program: Incident Command System Operational System Description, ICS–120–1, 1981. Available from: Operations Coordination Center, PO Box 55157, Riverside, Calif 92517. An extensive list of FIRESCOPE and ICS publications is available from this address.

Incident Command System, 1983, Available from: Fire Protection Publications, Oklahoma State University, Stillwater, Okla 74078-0118, (800) 654-4055, $13.00.

Incident Command System: Basic Orientation Course Training Package, 1982. Available from: California State Board of Fire Services, California Fire Service Training and Educational System, 7171 Bowling Dr, Suite 500, Sacramento, Calif 95823.

Law Enforcement Incident Command System (LEICS), 1985. Available from: Jerome Ringhofer, Deputy Chief, Desert and Mountain Command, San Bernardino County Sheriff's Department, PO Box 569, San Bernardino, Calif 92402.

Multi-Casualty Incident Operational Procedures Manual, 1986. Available from: California Fire Chiefs Association, 825 M St, Rio Linda, Calif 95673, $5.00.

National Interagency Incident Management System: Information and Guides, 1983. Available from: National Wildfire Coordinating Group, Publications Management System, Boise Interagency Fire Center, 3905 Vista Ave, Boise, Ida 83705, Free. An extensive NIIMS publications and forms list and prices are also available at this address.

WHERE TO GET INFORMATION ON ICS TRAINING

For information on ICS training, contact your local office of the U.S. Forest Service, your state forestry agency, or:

FIRESCOPE
Operations Coordination Center
P.O. Box 55157
Riverside, CA 92517

Director, Fire and Aviation Management
USDA Forest Service
PO Box 96090, Room 1001 RP-E
Washington, DC 20090-6090

U.S. Bureau of Land Management
Director, Boise Interagency Fire Center
Attention: Public Affairs Officer
3905 Vista Drive
Boise, ID 83705

TRIAGE

In contrast to most routine emergencies, efficient response in disasters requires procedures for triage and casualty distribution.
(Courtesy of California Office of Emergency Services, Sacramento, California.)

One aspect of resource management that deserves special attention is the distribution of resources for medical care. This chapter will address the concepts of triage and how they can be applied to affect medical resource management. The more common difficulties in disaster triage and some suggested solutions are discussed.

WHAT IS TRIAGE?

Traditionally, triage has been called the keystone to mass casualty management (Bowers, 1960:59). Triage comes from the French verb, *trier*, which means "to sort." It evolved, perhaps as early as Napoleon's time, as a technique for assigning priorities for treatment of the injured when resources were limited. The basic concept was *to do the greatest good for the greatest number of casualties.* Generally, attention is given first to those with the most urgent conditions and to those who are the most salvageable (Rund, 1981:3; Silverstein, 1984:8). The technique is considered by many to be essential for good disaster medical care (Spirgi, 1979:25; FEMA, 1983e:108; Cowley, 1982; Burkle, 1984:45).

> ## PRINCIPLE
> A basic concept of triage is to do the greatest good for the greatest number of casualties.

Doing the greatest good for the greatest number of disaster casualties does, however, involve more than just deciding who gets treated first. It also requires that use of all of the available treatment resources is maximized. That is, that the casualties are distributed rationally among the various hospitals and other medical treatment facilities (Silverstein, 1984:8, 44). Therefore, the definition of triage to be used in this text includes the organized evaluation of all disaster casualties to establish treatment and transport priorities. In addition, it involves the process by which casualties are rationally distributed among the available treatment facilities. Typically, management of triage is a *systems* problem requiring inter-organizational coordination and flow of information.

> ## PRINCIPLE
> Triage implies making the most efficient use of available resources.

There are three major reasons why triage is beneficial in the disaster response:

1) Triage separates out those who need rapid medical care to save life or limb.

2) By separating out the minor injuries, triage reduces the urgent burden on medical facilities and organizations. On average, only

10–15% of disaster casualties are serious enough to require overnight hospitalization.

3) By providing for the equitable and rational distribution of casualties among the available hospitals, triage reduces the burden on each to a manageable level, often even to "non-disaster" levels.

TRIAGE PROBLEMS IN DISASTERS

Observations in disasters have revealed problems with triage. The most comprehensive data collected to date are those from the Disaster Research Center obtained as part of a study of emergency medical services (EMS) in 29 major U.S. disasters occurring in the 1970s (Quarantelli, 1983; Golec, 1977). Because no similar studies have been carried out since then, it is difficult to determine to what extent these problems have been ameliorated by modern improvements in EMS systems and disaster planning. However, there is evidence to suggest that at least some of these problems continue to occur.

When Disaster Research Center investigators carried out these studies, they found an interesting discrepancy. In 55% of the cases studied, responders claimed that triage was carried out. However, the researchers found that the word "triage" was used in a loose fashion to describe almost any handling of the victims by emergency personnel. Sometimes the presence of uniformed medical personnel seemed to suggest to onlookers or other responders that triage was being carried out even when it was not. But if the term was used to describe appropriate assessment and sorting of all casualties according to the seriousness of their injuries, then little triage actually occurred (Quarantelli, 1983:69; Tierney, 1977:154).

Furthermore, in quite a number of disasters, casualties were not distributed among the available hospitals in a rational or efficient manner. Instead, the vast bulk of them ended up at the closest hospital, while other hospitals received no casualties at all. A variant of this pattern was where one hospital in the community was thought to give superior emergency care to critical casualties, or where it was more familiar to those providing the transportation. Such might be the case if one facility was renowned as the local "trauma center." In that event, the majority of victims sometimes ended up there (a pattern also observed in a number of previous disaster case studies) (Quarantelli, 1983:73; Golec, 1977; Rosow, 1977:166; Cohen, 1982a:19; Mileti, 1975:84; Williams, 1956:659; Neff, 1977:183).

In 75% of cases studied, a majority of the casualties were sent to the closest hospital. In 46% of the cases, more than three-fourths of the casualties were sent to the nearest hospital. Only in about half of the disasters did a simple majority of the hospitals in the area receive even one casualty. The pattern is illustrated by the figures in Table 8-1.

Table 8–1.   Hospital Distribution of Disaster Casualties

No. Casualties	No. Hospitals Receiving Casualties	No. Area Hospitals Capable of Receiving Casualties
266	4	43
141	4	41
381	12	78
298	11	105

(Adapted from: Quarantelli EL: Delivery of emergency medical services in disasters: Assumptions and realities, Irvington, New York, 1983, p. 88.)

Another perspective on the situation is given by Table 8-2 (Golec, 1977:171). (Note the percentage treated in one hospital.)

The hospitals *not* receiving patients had an average of 20% of their beds vacant (Quarantelli, 1983:79).

Not only did one hospital receive the largest number of casualties, but also those most seriously injured. In one disaster, for example, 40 out of 51 casualties were sent to one hospital which admitted 30 of them (28 in serious condition). The remaining 11 victims were taken to four other area hospitals. Not one of these 11 had injuries considered serious enough to require staying in the hospital. The pattern was similar for the casualties that were dead on arrival (Quarantelli, 1983:81). Even in those communities with only one hospital, a large community with a number of hospitals was usually located within 15 to 30 miles (Golec, 1977:172).

In considering the distribution patterns of disaster casualties, it should be noted that the optimal pattern does not necessarily mean that every hospital receives an equal number. In addition to hospital size and emergency department capacity, a facility's ability to take care of specialized cases (e.g., trauma) may affect the number of cases it can effectively handle. It could be argued that a trauma victim may be able to receive better care in a crowded trauma center than a less burdened but also less experienced facility. However, there is a lack of data on casualty severity versus receiving hospital capability in disasters, the level of care rendered, or the resulting mortality or morbidity. Therefore, the degree that overcrowding versus trauma experience affects patient outcome is yet to be determined.

In some disasters, it has been claimed that adequate care was given, even though hospitals received a disproportionate share of casualties (KC Health Dept, 1981:13, 16, 19; Ross, 1982:65; Moore, 1958:28; Lewis, 1980:863). Nev-

Table 8-2.   Distribution of Disaster Casualties

No. Treated	Hospitals Used/ Hospitals Avail.	% Treated in *one* Hospital
132	8/12	41%
34	4/11	65%
155	2/6	97%
28	2/3	89%
103	4/4	93%
140	4/17	90%
45	3/3	60%
94	2/7	60%
61	7/17	51%
55	4/5	55%
398	11/105	52%
35	4/105	71%
200	13/26	42%
51	5/12	78%
		Average (mean) 67%

[Deleted from this table are 3 communities with only 1 hospital]
(Adapted from Golec JA, Gurney PJ: The problem of needs assessment in the delivery of EMS, Mass Emergencies, 2:169–77, 1977.)

ertheless, reasonable balanced distribution of disaster casualties and especially the use of all hospitals to the levels of their capabilities, seems to be a reasonable measure of optimal medical resource use.

Despite the fact that the incidence and quality of triage has not been subjected to rigorous study since Quarantelli's research in the 1970s, anecdotal reports from recent disasters have revealed that under the right circumstances rather good patterns of casualty distribution to hospitals are possible. In some cases, use has been made of non-hospital treatment facilities for minor injuries, and casualties have been reasonably distributed among area hospitals.

> **EXAMPLE:** *High-rise Fire, MGM Grand Hotel, Las Vegas, Nevada, November 21, 1980.* The disaster plan called for the use of the Convention Center as a secondary triage and refuge center. This facility, remaining from earlier civil defense planning, was equipped as an acute care hospital and contained 200 beds and 100 additional cots. It was staffed primarily by Red Cross

and other volunteer staff. Although it was not used as a hospital in this disaster, 1,700 minimally injured and displaced persons out of the 6,000 hotel guests were triaged to the center, many of these by bus. Of 769 injured survivors, 150 received treatment by medical teams at that location. Others were transported for treatment to the following facilities:

Southern Nevada Hospital—104

Desert Springs Hospital—161

Sunrise Hospital—211

Valley Hospital—143 (Buerk, 1982:641; Morris, 1981:20).

EXAMPLE: *DC-9 Airline Crash, Stapleton International Airport, Denver, Colorado, November 16, 1987.* Nine of the area hospitals usually accept emergency ambulance patients on a daily basis. In this disaster, 10 hospitals received the following numbers of the 56 injured survivors and 5 injured rescuers:

Denver General Hospital (a level I trauma center)—3 immediate, 24 minor;

University Hospital (a level I trauma center)—6 immediate, 2 minor;

St. Anthony, Central (a level I trauma center)—3 critical, 1 delayed, 2 minor;

Swedish Hospital (a level II trauma center)—1 immediate, 1 delayed, 1 minor;

Presbyterian Aurora Hospital—2 delayed;

St. Joseph Hospital—1 immediate, 1 delayed, 3 minor;

St. Luke Hospital—2 delayed;

Porter Memorial Hospital—1 delayed, 2 minor;

Fitzimons Army Hospital—1 delayed, 1 minor;

Rose Medical Center—2 delayed, 1 minor (Dinerman, 1988).

It should be noted, however, that in each one of these cases, the disaster covered a rather small geographic area. Adequate triage and casualty distribution is more difficult to achieve in disasters such as tornadoes, floods, hurricanes, and earthquakes, that cause injury and destruction over a wide area.

PRINCIPLE
Good casualty distribution is particularly difficult to achieve in "diffuse" disasters, such as earthquakes and tornadoes, that cover large geographic areas.

As shown in Table 8-3, observations in a number of cases reveal that communities did not take full advantage of all their available hospital resources in disasters.

Table 8–3.   Distribution of Casualties

Disaster	Distribution Pattern
Tornado Flint-Beecher, MI 1953	About 750 (80%) of the victims brought to one hospital. (Rosow, 1977:165)
Tornado Worcester, MA 1953	800 injured; 90% admitted by 3 of 15 hospitals. (Rosow, 1977:63,104)
Tornado Waco, TX 1953	Most of the injured were taken to one downtown hospital, which was swamped before those at greater distance were filled. (Moore, 1958:23)
Aircraft carrier fire NY 1960	A nearby hospital had 35 patients and inadequate resources to care for them. Most could have been transported to hospitals farther from the scene where adequate personnel were available and provisions for treatment had been made. (Shaftan, 1962:113)
Coliseum explosion Indianapolis, IN 1963	75% of the 374 casualties taken to 3 hospitals in town. Hospitals available included 7 in the city and 12 in surrounding area, with total beds numbering in excess of 4,600. (Drabek, 1968:23)
Earthquake Alaska 1964	Nearly all of the victims taken to 1 of 5 hospitals in Anchorage. (Yutzy, 1969:64)

cont'd.

Table 8–3.   Distribution of Casualties *cont'd.*

Disaster	Distribution Pattern
Train crash Chicago, IL 1972	Of the 400 injured, none were transported to Cook County Hospital, the local trauma center, a 4-minute trip by any of the 15 helicopters that were transporting patients from the scene. (Cihlar, 1972:17b)
Train crash Chicago, IL 1976	Of 381 injured, 85% sent to 3 out of 11 hospitals, none to Cook County Hospital. (Mesnick, 1980:134)
Train crash Chicago, IL 1977	Of 183 injured, 48% sent to 2 closest out of 11 hospitals; Cook County Hospital received 9%. (Mesnick, 1980:136)
Hyatt Hotel Skywalk Collapse Kansas City, MO 1981	17 of 26 hospitals used; 4 closest hospitals received 42% of the 200 victims and 55% of those admitted, and did 83% of the surgery. (KC Health Dept, 1981:12,13,15)
Air Florida crash Washington, DC 1982	19 of 22 injured (86%) went to one hospital. (Edelstein, 1982:159)

CAUSES OF TRIAGE PROBLEMS

Responders from Outside the Local EMS System

Non-ambulance Transport of Casualties

One of the difficulties that faces emergency medical services (EMS) systems trying to carry out triage is that many injured casualties reach the hospital outside the EMS system (Quarantelli, 1983:63,70; Tierney, 1977:155; Mileti, 1975:84; Golec, 1977:175; Seismic Safety Comm, 1983:83; Scanlon, 1988:6). Table 8-4 indicates the means of arrival of the *first* casualties at 75 hospitals where the method of transport could be determined.

Table 8–4. Means of Initial Disaster Casualty Arrival at the Hospital

Ambulance	54%
Private Auto	16%
Police Vehicle	16%
Helicopter	5%
Bus or Taxi	5%
On Foot	4%

(Adapted from: Quarantelli EL: Delivery of emergency medical services in disasters: Assumptions and realities, Irvington, New York, 1983, p. 70.)

While these figures indicate the mode of arrival of the *initial* casualties, overall, less than half arrived by properly equipped ambulance or rescue vehicle. The researchers noted a strong tendency for police officers to load victims into whatever vehicles were handy and send them off to the hospital. In one disaster, for example, police loaded 26 injured persons into three non-ambulance vehicles, and these were the first to arrive at the hospital (Quarantelli, 1983:70).

Some disaster plans call for a delay in evacuation of victims from the scene. This is so they can be triaged for orderly and rational field stabilization and transport. Other plans call for the use of field first-aid stations to alleviate the burden on hospitals. However, these plans do not always take into consideration the perceptions and motivations of the victims or the public, which may be different from those of the planners. Failure to do so results in plans which may look good on paper, but which do not correspond to reality.

Often the *public's* perception of good emergency medical care is transportation to the hospital as quickly as possible (Drabek, 1968:148; Quarantelli, 1983:72,110; Quarantelli, 1970a:383; Wright, 1976:27). If medical care and transportation are not furnished promptly by official emergency organizations, victims do not usually sit idly by and await its arrival. Instead, they get themselves to the hospital by the most expedient means available. Often, they will go to the nearest hospital, the one with which they are most familiar, or the one in which they have the greatest trust. Field disaster first-aid stations are often bypassed, either because their location is unknown, or because for many people "first aid" is seen as an inferior level of medical care. This pattern has been noted most particularly in diffuse, widespread disasters such as tornadoes and earthquakes (Wright, 1976:27; Quarantelli, 1970a:384; Dynes, 1974:30; Quarantelli, 1983:21,64; Raker, 1956:23; Drabek, 1986:139,170; Adams, 1981b:17,30,57; Worth, 1977:161).

EXAMPLE: *Earthquake, Coalinga, California, May 2, 1983.*
Only 7 of the 38 casualties arriving at the Coalinga District
Hospital in the first hour came by ambulance. The rest came by
private car or on foot. The most seriously injured victim arrived
in the vehicle of a local physician. Another local doctor, who
was responsible for the city's disaster medical response, established a triage site in the devastated downtown area. All patients from the downtown area, however, went directly to the
hospital, bypassing the triage site (Seismic Safety Comm,
1983:83; Kallsen, 1983:25).

EXAMPLE: *Tornado, Edmonton, Alberta, Canada, July 31,
1987.* Out of more than 300 injured victims, 30% were transported to the hospital by a family member, 20% were taken by a
stranger, 18% arrived by bus, and 16% were conveyed by
ambulance (Scanlon, 1988).

OBSERVATION
The distribution of disaster casualties is complicated by the tendency for the victims to get *themselves* to the nearest hospital or the
one with which they are most familiar and in which they have the
most trust.

This rapid transport to nearby hospitals by non-ambulance vehicles contributes to two problems seen frequently in disaster situations: 1) Casualties with
relatively minor injuries arrive (often unannounced) before those with serious
conditions. The result is that when the more serious victims arrive, the hospital
emergency department is already inundated and its beds occupied. 2) Casualties arrive at the hospital without having been triaged or having received
stabilizing first aid (Quarantelli, 1983:73).

Involvement of Non-local Responders
Another factor that has contributed to the lack of organized triage and casualty
distribution has been the number of responders from non-local organizations
and those not under control of the local EMS system. This has been true
especially in larger disasters and those occurring in urban areas (Quarantelli,
1983:68,71; Morris, 1982a:65; Seismic Safety Comm, 1983:81,86. The increased
use of helicopters for medical transport seems to contributed to this trend
(seismic safety comm, 1983:81,86; Quarantelli, 1983:70).

EXAMPLE: *Kansas City Hyatt Hotel Skywalk collapse.* The City Health Department post-disaster review noted that coordination was never established over patients transported by a helicopter operated by one of the local hospitals. The crew reportedly failed to coordinate with those directing scene activities, including the ambulance dispatcher, the triage officer, or the site communications officer (KC Health Dept, 1981:7).

Effects of Search and Rescue Activities

The manner in which search and rescue activities are carried out has an important influence on triage. Search and rescue often becomes the initial contact point with the disaster victims. Therefore, those carrying out this activity generally influence how the disaster casualties enter the EMS system. When search and rescue operations are confused and uncoordinated, the flow of casualties into the EMS system tends to develop the same characteristics (Quarantelli, 1983:63,67).

Search and rescue is not always coordinated or carried out with significant input by those having emergency medical expertise (Quarantelli, 1983:66). In part, this is due to the large amount of search and rescue that is carried out by unofficial civilian volunteers, often family members and neighbors (Drabek, 1981:xviii,38,53,68,87,97,111,119). It is also due to the ambiguity regarding who has the overall responsibility to coordinate search and rescue operations (Quarantelli, 1983:67; Drabek, 1981:xx,35,240; Wenger, 1986:32).

Lack of Inter-organizational Planning

In many disasters, the flow of disaster casualties into the EMS system has not occurred according to any formal, pre-designated plan. In the 1970s, the Disaster Research Center found casualty flow occurring according to plan in only about half the cases they studied. In part, this occurred because many communities did not have a realistic inter-organizational plan for disaster EMS. This was the case in over 66% of the localities studied. Furthermore, even when plans existed, they were often limited in scope, dealing with only a single jurisdiction, or calling for the coordination of only 2 or 3 of the community's emergency agencies. In only about 25% of the cases studied was there anything resembling a *region-wide* plan (Quarantelli, 1983:71,86,101,103,106). Although there have been improvements in disaster planning since the 1970s, difficulties continue to be seen (Wenger, 1986:ii).

Even when plans for triage exist, they may be "paper plans." That is, they are either unrehearsed, devoid of associated training, based on invalid assumptions, or encompass only a limited number of those who actually participate in the disaster response (Dynes, 1981:75; Quarantelli, 1985:7).

Lack of Needs Assessment

Efficient use of available medical resources (including hospital facilities) requires an overall needs assessment to determine the numbers, types, and severities of injuries. It is also crucial to ascertain the availability and status (available versus in-use) of medical resources such as field medical personnel (EMTs, paramedics, nurses, physicians); equipment; ambulances and rescue and/or first-aid vehicles; and hospital facilities. However, disaster response has often evolved without consideration of the overall situation. Rather, many individuals did what seemed rational from their own isolated perspective. That is, they endeavored to move each disaster victim to the closest hospital as quickly as possible. Even when triage did occur, most often it included only an assessment of specific individual casualties, rather than an evaluation of the disaster as a whole (Quarantelli, 1983:111; Golec, 1977:169).

When the medical aspects of situation analysis have been neglected, two factors in particular seem to have contributed to this oversight.

Lack of Medical Direction at the Scene

In some cases, those with emergency medical training have not played a major role in the overall direction of activities at the disaster scene. Others likely to be directing disaster site operations may lack familiarity with the function of emergency medical systems (EMS) (Quarantelli, 1983:66; U.S. Fire Admin, 1980:3,18,21,28,41,43).

Lack of Scene-to-Hospital Communications

Communications between the scene and area hospitals is essential for situation analysis and casualty distribution, yet meaningful and informative scene-to-hospital information flow (see Chapter 5) is often neglected (Golec, 1977:174; Neff, 1977:186; Edelstein, 1982:159; Goodwin, 1982:14). The Disaster Research Center found less than 22% of the cases where meaningful information exchange occurred between the disaster site and any area hospital (Quarantelli, 1983:67). Even the existence of elaborate, pre-planned procedures for scene-to-hospital communications related to triage activities, was no guarantee that the procedures would actually be used.

> EXAMPLE: In one community, the disaster plan included procedures to prevent the overloading of any single hospital. The central communications center had access to information

on each hospital's bed census and emergency department capability. The communications center was to notify the hospitals in the event of a disaster and was to direct patients away from overloaded hospitals. In spite of the plan, 90% of the 140 casualties were taken to one hospital out of 17 in the community. The remaining 15 were distributed among three other hospitals. Furthermore, the communications center never even notified the hospital that the disaster had occurred (Golec, 1977:172).

IMPROVING TRIAGE

Many of the general principles applicable to disaster management in general may improve triage. Use of common terminology and the existence of joint planning, training, and testing all contribute to effective activity. Procedures for cooperative communications, situation assessment, resource management, and integration of unexpected or unfamiliar responders are all applicable to organized triage efforts.

Coordination with Non-Medical Organizations

Successful triage is dependent not only on the actions of medical (EMS) personnel at the site, but on non-medical responders as well. Often, the majority of casualties in disasters are initially encountered during search and rescue efforts. Although the very first search and rescue is usually carried out by civilians who happen to be in the impact area, when the activity is taken over by formal emergency responders, they are most likely to be firefighters or peace officers (Quarantelli, 1983:66). Triage is more successful if injured casualties located by search and rescue efforts are fed into the triage system. This requires a concerted effort by those overseeing these two essential activities.

Coordination is also important with other organizations whose activity might affect triage. Examples are those responsible for adequate crowd and traffic control, decontamination of those exposed to hazardous substances, and provision of light and shelter for triage areas.

Coordination with Hospitals

Notification of Hospitals

Functional procedures are required to designate a person whose responsibility it is: 1) to see that all area hospitals are notified that a disaster exists and provided with information regarding its location, character, magnitude, and the numbers, types, and severities of casualties to expect; 2) to continually and regularly update this information; 3) to respond to requests from the hospitals for further information; and 4) to indicate when the hospitals may deactivate their disaster status.

Hospital Capacity Assessment

In order to distribute casualties rationally among area hospitals, someone at the scene needs to be responsible for acquiring information from the hospitals regarding their capacities and capabilities. This information needs to be updated continually, because hospitals are also likely to be receiving casualties who have gotten there by their own means. In addition, as off-duty staff come in, the hospital may be able to care for more patients than when the facility was initially notified. In some communities, one of the local hospitals is designated as a "disaster coordination hospital," responsible for collecting capacity information from the hospitals and casualty information from the scene. This facility is then responsible for directing ambulance destinations based on this information.

PRINCIPLE
Effective triage requires coordination among medical and non-medical organizations at the disaster site and between the site and local hospitals.

Coordination of Scene Medical Activities

One model for the coordination of scene medical activities is that described in the 1986 version of the California Fire Chief's Association, Multi-Casualty Operational Procedures (MCOP) Manual (CFCA, 1986). What follows is a brief description of the system. For more detailed information, a copy of the manual is available from the Association at 825 M Street, Rio Linda, CA 95673, $6.50.

The system is designed as a medical component of the Incident Command System (ICS) (see Chapter 7), and uses procedures and terminology consistent with ICS. The organizational structure is diagrammed in Fig. 8-1.

This structure may be expanded to encompass multiple triage areas as illustrated by the diagram in Fig. 8-2.

Each position is provided with a checklist of responsibilities similar to that type of checklist used for the ICS. An example is given in Fig. 8-3.

Similar checklists are provided for all positions in the Multi-Casualty Incident Procedures Manual.

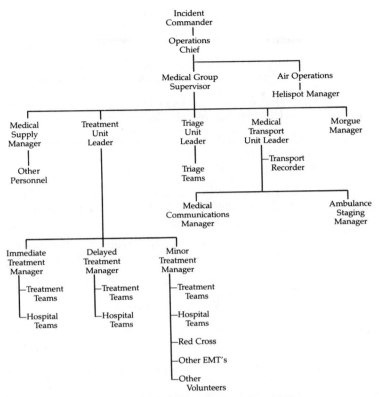

Figure 8-1.   Organizational structure of triage. (Adapted from Multi-casualty incident operations procedures manual, Rio Linda, 1986, California Fire Chiefs Association.)

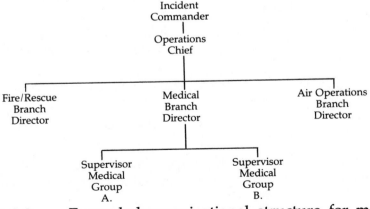

Figure 8-2.   Expanded organizational structure for multiple triage sites.

MEDICAL GROUP SUPERVISOR

Definition: Qualified officer.

Commanded by: Division Supervisor, or Branch Director, or Operations Chief, or Incident Commander [whichever is the lowest position that is activated].

Subordinates: Triage Unit Leader, Treatment Unit Leader, Medical Transportation Unit Leader, Medical Supply Manager, Morgue Manager.

Function: Establish, command, and control the activities within a Medical Group in order to assure the best possible emergency medical care to patients during a multi-casualty incident.

Duties: 1. Establish and supervise a Medical Group at a level of personnel and other resources sufficient to handle the magnitude of the incident.

2. Delineate officers and designate patient control area locations as appropriate. Isolate minor treatment and morgue areas.

3. Ensure law enforcement/coroner involvement as necessary.

4. Ensure activation of hospital alert system.

5. Request Hospital Emergency Response Teams through the hospital alert system as necessary to provide medical assistance.

6. Determine amount and types of additional medical resources and supplies, e.g., Medical Strike Teams, Medical Task Forces, medical caches, ambulances, helicopters, and other methods of patient transportation.

7. Establish coordination of air ambulance (helicopter) operation between Medical Transportation Unit Leader and the Air Operations Director.

8. Establish liaisons with on-scene agencies, e.g., Coroner's Office, Red Cross, law enforcement, ambulance companies, county health agencies, etc.

9. Ensure that proper security, traffic control and access have been established.

10. Direct other medically trained personnel to appropriate unit leaders.

Figure 8-3.   Medical group supervisor duty checklist. (Adapted from Multi-casualty incident operations procedures manual, Rio Linda, 1986, California Fire Chiefs Association.)

The flow of casualties and layout of triage, treatment, and transportation areas are illustrated in Figs. 8-4 and 8-5.

TRIAGE PROCEDURES

Examples of Triage Classification Systems

There is no single, standard, or universal method of triage. The number of categories used may vary from 2 to 5 or more, depending on the particular system in use. Various color codes, numbers, and symbols have been used to identify these categories. The triage category is often identified by the use of a triage tag, the design of which is also variable. In the absence of a triage tag, a triage symbol is sometimes written on the patient. The selection of how many categories or what colors or symbols to use for triage is somewhat arbitrary, and each system has its particular advantages and disadvantages. If the number of

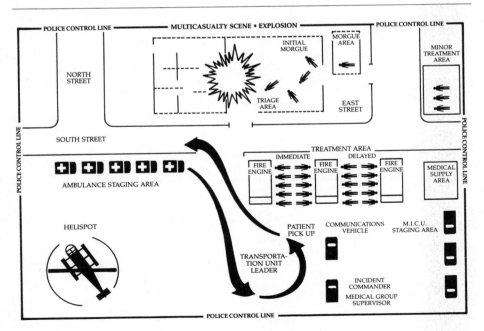

Figure 8-4. Multi-casualty scene. (Adapted from Multi-casualty incident operations manual, Rio Linda, 1986, California Fire Chiefs Association.)

Patient Flow Chart

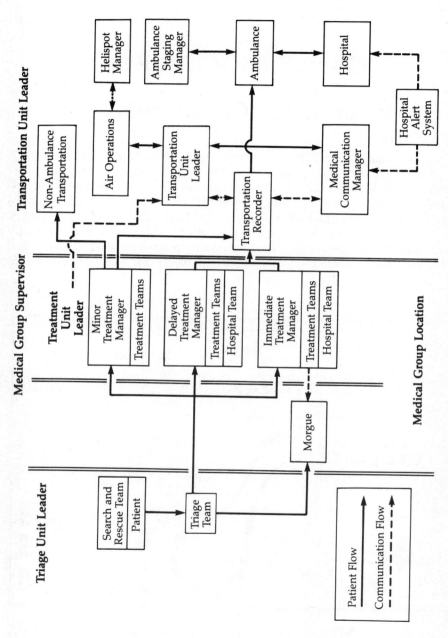

Figure 8-5.   Patient flow chart. (Adapted from Multi-casualty incident operations manual, Rio Linda, 1986, California Fire Chiefs Association.)

categories is limited to two, for example, the system is simple to remember. On the other hand, the use of more categories has the advantage of greater precision (Rund, 1981; Savage, 1977; Gazzaniga, 1979; Baker, 1979; Moore, 1967; Grant, 1982; Silverstein, 1984:12,63; Cohen, 1982a).

In order to illustrate how triage categories can be used, examples of two classification schemes will be given. The idea is not to endorse any particular system, but to illustrate a small sample of the various methods which demonstrate the basic concepts of triage.

The S.T.A.R.T. System

START is a mnemonic for Simple Triage And Rapid Treatment. This program was developed in southern California by a group of emergency physicians, firefighters, and an emergency nurse (Super, 1984).

The basic process for determining categories is represented by Fig. 8-6.

Information and training materials for the START system may be obtained from:

S.T.A.R.T.
Hoag Memorial Hospital Presbyterian
301 Newport Blvd., Box Y
Newport Beach, CA 92663
(714) 760-5689

A Proposed 5-Category Triage System

An alternative proposed triage system is described below. This system was designed to address some problems associated with the triage and management of "unsalvageable" disaster casualties. It was also constructed to be adapted not only to disasters, but also daily EMS care. The daily use of routine triage for EMS helps to assure that it will be a familiar system when disaster strikes. This system uses five color-coded priority categories. These are felt to be a representative sample of what is in common use (Savage, 1977; Baker, 1979). However, other effective systems use three or four categories. A summary of the system is given in Table 8-5, and its special features are then discussed.

Used in this context, "simple care" is that which does not require unusual expenditures of time, equipment, or personnel. "Simple" field care might include inserting an airway, sealing a penetrating chest wound, applying a MAST unit, or giving intravenous fluids for shock. "Complicated" care would include artificial ventilation or CPR. "Simple" hospital care might include giving intravenous drugs, applying splints, surgical cleansing of flesh wounds, performing a cricothyrotomy or inserting a chest tube. In some cases, it might also include surgical exploration of the abdomen. It would not include repair of

Step 1

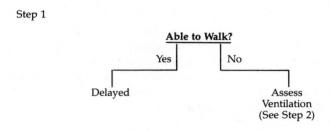

Step 2

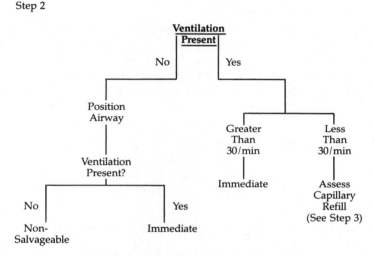

Figure 8-6. The START classification protocol. (Adapted from Super, G.: START instructor's manual, Newport Beach, Presbyterian Hoag Memorial Hospital.)

a transected aorta or ruptured aortic aneurysm, nor would it include surgery for a fractured neck.

The "catastrophic" category deserves special discussion. In some triage systems this has been called the **Expectant** category. It was reserved for those who were moribund or who were in such poor condition that they could only be saved if extensive resources were diverted from more salvageable cases. Immediate care was to be withheld from these cases so that limited resources could be used to do the most good for the most casualties. These casualties were grouped with the dead and minor casualties and given last priority (Tintinalli,

Step 3

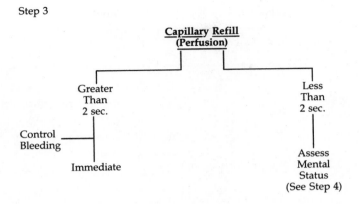

Step 4

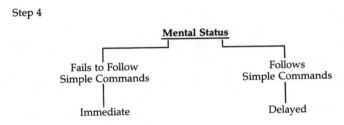

Figure 8-6.   *cont'd.*

1978). This assignment of last priority was probably a holdover from the days when much of the military casualty sorting was carried out at the battalion aid stations where resources were extremely limited (Rund, 1981). It had to be assumed under these conditions that these casualties would not survive. In the Vietnam War, however, helicopters evacuating casualties often bypassed the battalion aid stations and delivered the casualties directly to the hospital. At the hospital, *all* living casualties were initially considered potentially salvageable (US Dept Defense,1975; Rund,1981). This may be a more appropriate procedure for triage in the civilian setting.

Table 8-5.   5-Category Triage System

PRIORITY	COLOR	SYMBOL	CASUALTY CONDITION
FIRST	RED	R	CRITICAL: likely to survive if simple* care given within minutes.
SECOND**	BLUE	B	CATASTROPHIC: Unlikely to survive and/or extensive or complicated care needed within minutes.
THIRD	YELLOW	Y	URGENT: Likely to survive if simple* care given within hours.
FOURTH	GREEN	G	MINOR: likely to survive even if care delayed hours to days. May be walking OR stretcher cases.
NONE	BLACK	X***	DEAD

*Simple: Care that doesn't require unusual equipment, or excessive use of time or personnel.

**Assigned THIRD priority (after YELLOWS) when there are so many casualties that if resources are used in vain to try to save BLUE cases, the YELLOWS will needlessly die.

***The circling of this symbol prevents its being confused with a sloppily written Y.

Assigning last priority to **Catastrophic** cases and reserving that category for those with inevitably fatal conditions presents several problems:

- It is generally well adapted only for use in situations where there are truly *massive* numbers of casualties and access to extremely limited resources. These conditions are not at all typical of civilian disasters in the United States.

- Psychologically, it is untenable to "condemn" living casualties to such a category, especially under conditions of maximum duress, and by persons not experienced in making such decisions (Gann, 1979; Gazzaniga, 1979; Grant, 1982; Moore, 1967; Tintinalli, 1978). Fortunately, it is unlikely for any one person to encounter civilian mass-casualty situations frequently enough to become "experienced."

- Treating these patients last means that minor injuries (which may survive for days without treatment) are given priority over casualties that *may* still be salvageable once the resources needed to care for them are finally mustered.

Assigning last priority to catastrophic casualties, therefore, is probably not a realistic approach in the civilian setting. In addition, it does not work well if the triage system is to be used on a daily basis for routine emergency patients.

The approach described here, therefore, is to first treat the Critical (RED) casualties. After this, the Catastrophic (BLUE) casualties, which it may still be possible to salvage, are usually treated. When there are large numbers of casualties, it is conceivable that those initially categorized as Urgent (YELLOW) will eventually reach a point where they will need care within minutes. If it appears that this will occur before the Catastrophic (BLUE) casualties have all been treated, then attention is diverted from the BLUEs to first treat the YELLOWs. Minor (GREEN) casualties are not treated until attention has first been given to REDs, YELLOWs, and BLUEs.

Detailed examples of the patient problems in each category are given in Appendix D.

Triage Tags

It is commonly suggested that disaster casualties have their priority indicated by the attachment of a triage tag. There is no universal agreement regarding the design of such tags. Several useful variations are in use (Cohen, 1983; 1977; Cohen, 1986).

The S.T.A.R.T. System uses a commercial triage tag (METTAG) with four categories, indicated by four tear-off, colored strips at the bottom of the METTAG (see Table 8-6).

Table 8-6. The Triage Classification System Used by METTAG

GREEN: (Bottom strip)
    Symbol: ambulance—crossed-out
    Meaning: No hospital treatment needed; first aid only

YELLOW: (Second strip from bottom)
    Symbol: Turtle
    Meaning: Non-urgent; hospital care

RED: (Third strip from bottom)
    Symbol: Rabbit
    Meaning: Urgent; hospital care

BLACK: (Fourth strip from bottom)
    Symbol: Cross/dagger
    Meaning: Dead or unsalvageable; no CPR

(Adapted from: METTAG literature, Starke, FL)

There are several practical features of this tag. It is designed so that if a casualty's condition deteriorates, the next strip can be torn off to indicate this fact. Each tag has an identification number on it and on each of the colored strips. In addition, two upper corners of the tag have the number on it, and they can be removed by tearing along perforations, and used for keeping track of the casualties. The upper part of the tag has spaces for patient information. Two disadvantages have been noted with the METTAG: 1) some responders have complained that the colored strips on the METTAG are hard to see at a distance; and, 2) whereas a patient's deteriorating condition can be indicated on the METTAG by tearing off an additional strip, an improving condition cannot be so easily indicated.

The amount and type of casualty information to be placed on the triage tag is also by no means standardized. The tag may provide for notations regarding such things as locations and types of injuries (sometimes indicated with a diagram of the human body), pulse, respiration, blood pressure, treatment given, a serial number, and the patient's name, address, age, gender, and next of kin.

Another approach is used by the Alpine, Mother Lode, and San Joaquin Emergency Medical Services Agency in California. They use a tag for patient information, and a separate, colored cloth tag to indicate the triage category. Thus, when the casualty's condition improves or deteriorates, the cloth tag is changed and the information tag remains with the patient. These cloth tags are inexpensive, durable, and visible from a distance.

One problem that has been observed in disasters is that, while the disaster plan called for the use of triage tags, there was a lack of tags at the incident site (KC Health Dept, 1981:5; Buerk, 1982:643; Quarantelli, 1983:77; Worth, 1977:163). The best solution to this problem is to keep a set of triage tags aboard every emergency rescue and ambulance vehicle. Each batch of tags should also include a summary of triage plan and categorization scheme, as well as charts for keeping track of casualty conditions, hospital capabilities, and hospital destinations.

Casualty Distribution Procedures
Variations Depending on Local Conditions
There are a number of different approaches to disaster casualty distribution. Which approach is most practical may depend on the size of the community, the number of area hospitals, and the difference in capabilities of these institutions.

In the simplest case, only one local hospital is available. However, it may be necessary for this hospital to act as a triaging facility, stabilizing patients then distributing them to more distant facilities (Butman, 1982:140). If there are but a few hospitals in the community, all with similar capabilities, it might suffice to send one ambulance load to each facility on a rotating basis. In large urban centers, a rather sophisticated set of distribution procedures may be necessary.

Hospital Polling

In some communities, the disaster plan may include procedures for polling each hospital to obtain information about its present staffing, number of empty beds, operating room availability, and other resources. Such plans need to recognize the time it takes to collect this information and to consider the fact that casualties typically begin to *arrive* at the hospital within 30 minutes of disaster impact (Quarantelli, 1983:74; Golec, 1977:173). In order to be of maximum use, the polling information has to be made available to field medical units before the casualties leave the scene.

The "First-Wave" Protocol

In communities that use hospital polling, it may be advantageous to predetermine a method for equitably distributing the initial disaster casualties, pending the collection of hospital information. By its very nature, such a procedure is likely to be imprecise. Nonetheless, even a fairly crude distribution of casualties is better than what is often achieved if no procedure is in effect.

An example of such a procedure is the first-wave protocol. This involves the predetermination of disaster treatment capabilities of the area hospitals as a *guideline* for casualty distribution. These are based on the "worst-case" types and numbers of casualties each facility can treat (i.e., 2 a.m. on a Saturday). These categories are established to match those used in the local triage categorization system. Using the above five-tiered triage system as an illustration, the hospital capable of treating a minimum of three urgent ("yellow") casualties would be designated as a "yellow-3" first-wave facility. Another hospital, capable of treating only minor casualties, but which could manage 20 of them, would be designated a "green-20" first-wave facility and so on. Although priorities are different for "red" (critical) and "blue" (catastrophic) casualties, the facilities needed for treatment are the same. Therefore, red and blue casualties would be sent to facilities with a "red" designation. Communities may consider the use of guidelines established by the Committee on Trauma of the American College of Surgeons in the selection of facilities for critically injured casualties (ACS, 1986:4).

Using this protocol, a community distributes casualties according to the type and number in the first-wave designation of each facility. When, for example, a "red-3" facility has received three critical casualties, any further critical victims are sent to other "red" facilities until all such hospitals have received their quota. A modification of this system (applied only to critical casualties) has been initiated in Sacramento County, California (Lowry, 1983).

When all hospitals of any particular color designation have received their share of casualties, subsequent distribution is according to a calculated first-wave ratio. This is determined by adding up all the hospital capacities for a triage category and dividing by the number indicating the capacity of each. For example, if the community has a total of 10 "red" casualties, and Hospital A has a first-wave designation of 4, then 4/10 or 40% of the "red" casualties from the disaster are sent to this facility. First-wave designations for a hypothetical community are illustrated in Table 8-7.

The application of this First-Wave protocol is illustrated in Fig. 8-7.

Unfortunately, the existence of a disaster does not necessarily diminish the occurrence of routine emergencies. People continue to have babies, get sick, get drunk, and crash their cars into each other. The first-wave protocol can be used to take into consideration all the accidents and illnesses occurring in the community, including those created by the disaster. In this case, the sum of all the critical casualties at the disaster site and occurring in other areas of the community are used to determine the total load of critical casualties to be distributed to "red" hospitals. In a similar manner, distribution is determined for other categories of patients both on and off the disaster scene.

As in the case of triage categories, distribution techniques can be adapted for use in routine emergencies. When any hospital in the community, as the result of one or several emergencies, receives simultaneously a total number of emergency patients exceeding its first-wave score, this can be used as a guideline for considering temporarily redirecting ambulance traffic to other facilities. Likewise, when all "red" facilities have patient loads exceeding their first-wave scores, then the first-wave ratios can be used as a guide to divvy up the patient load. The adaptation of such types of disaster procedures for more routine emergency situations, tends to assure that the users keep familiar with them.

SUMMARY

Triage, a wartime invention, involves the concept of "doing the most good for the most casualties." As such, it is well adapted for use in civilian disasters. Triage is often thought of in narrow terms as merely the designation of priorities for patient care. However, in disasters, doing the most good for the most casualties also means maximizing the use of the available hospital facilities. This is often a difficult task to accomplish, especially in diffuse disasters covering a large geographic area. This chapter has examined some of the more common problems interfering with effective triage and casualty distribution in disasters. The reasons for these problems and some examples of how to counter them have been discussed.

Table 8-7. Calculating the First-Wave Score and Ratio

A	B	C	D	E
Designation	Hospital	First-Wave Score	Total First-Wave Score of All Area Hospitals*	First-Wave Ratio
Red	Hospital A	4	10	4/10 = 40%
Red	Hospital B	2	10	2/10 = 20%
Red	Hospital C	2	10	2/10 = 20%
Red	Hospital D	2	10	2/10 = 20%
Yellow	Hospital E	8	20	8/20 = 40%
Yellow	Hospital F	12	20	12/20 = 60%
Green	Hospital G	5	20	5/20 = 25%
Green	Hospital H	15	20	15/20 = 75%

*with the same "designation"

Determine the highest triage category that the hospital can manage at any time of day or week, giving good disaster care for the initial 2 hours, without calling in disaster back-up resources. Place this designation in column A.

Determine the maximum number of casualties that can be managed at one time under such conditions. Place this number in column C.

Determine the *total* number of casualties of this triage category that can be managed by all hospitals in the area. Place this number in column D.

Divide the number in C by that in D (multiply the product by 100 to convert it to a percentage). Place this ratio in column E.

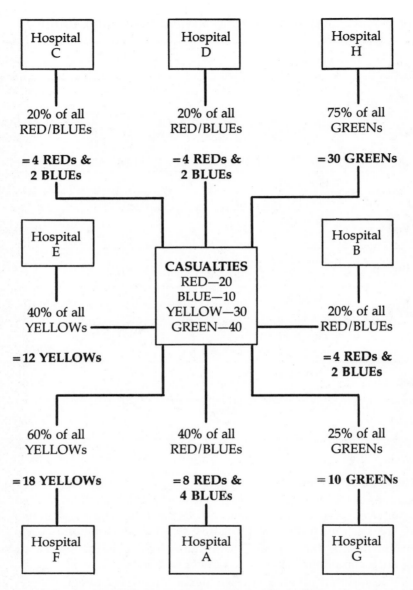

Figure 8-7.   Distribution using the "First-Wave" protocol.

PLANNING CHECKPOINTS

[ ] Do persons with expertise in emergency medical services have primary authority over patient care and transport at the scene of a disaster?

[ ] Do your fire and police personnel understand that they may be involved in disaster search and rescue operations?

[ ] Do they understand how they should coordinate their activities with those involved in triage at the site?

[ ] Does the disaster plan and training provide for close contact between those directing search and rescue and those providing EMS?

[ ] Does your community have a plan and associated training for disaster casualty distribution among area hospitals?

[ ] Are your disaster triage and distribution procedures adapted so they can be used in more common emergencies?

[ ] Does your community have procedures and training for assessing the types, numbers, and severities of casualties at the scene? For sharing this information with all the involved responding organizations including all area hospitals?

[ ] Is this information collected and disseminated on an *ongoing* basis?

[ ] Does your community have functioning procedures and training for assessing the capacity and capability of local hospitals?

[ ] Does every ambulance and rescue vehicle have a supply of triage tags?

[ ] Does your plan anticipate that "trauma centers" and hospitals near the disaster tend to get a disproportionate share of casualties?

[ ] Are all public safety and EMS personnel required to have ongoing training familiarizing them with the local disaster triage and EMS system?

[ ] Does your plan and training indicate who is responsible for each of the following field disaster medical care/triage responsibilities:
    [ ] Overall coordination?
    [ ] Liaison with other agencies?
    [ ] Dispatch?
    [ ] Provision of disaster information to hospitals?
    [ ] Assessment of up-to-date hospital capabilities?
    [ ] Casualty distribution?
    [ ] Triage?
    [ ] Patient care at the scene?
    [ ] Logistics?

[   ] Air transport?
[   ] Public information?

ADDITIONAL READING

Burkle FM Jr, Sanner PH, and Wolcott BW: Disaster medicine: application for the immediate management and triage of civilian and military disaster victims, 1984. Available from: Medical Examination Publishing Co, Inc, 3003 New Hyde Park Rd, New Hyde Park, NY.

Butman AM: Responding to the mass casualty incident: a guide for EMS personnel, 1982. Available from: Emergency Training, 181 Post Road West, Westport, Conn 06880.

Cohen E: A better mousetrap: what makes up the "perfect" triage tag?, J Emerg Med Serv, pp. 30–36, July 1983.

Cohen E: Patient identification: a look at triage tags, Emerg Med Serv, 15(9):45–49, Oct 1986.

Eisman B: Combat casualty management in Vietnam, J Trauma, 7(1):53–63, 1967.

Emergency War Surgery: First United States revision of the emergency war surgery NATO handbook, United States Department of Defense, 1975. Available from: Superintendent of Documents, US Government Printing Office, Washington, DC 20402.

Golec JA and Gurney PJ: The problem of needs assessment in the delivery of EMS, Mass Emergencies, 2:169–177, 1977.

Multi-Casualty Incident Operational Procedures Manual, 1986. Available from: California Fire Chief's Association, 825 M Street, Rio Linda, Calif 65673.

Quarantelli EL, Delivery of emergency medical services in disasters: assumptions and realities, 1983. Available from Irvington Publishers, 551 Fifth Avenue, New York, NY 10017.

Rund DA and Rausch TS: Triage, The C V Mosby Co, St Louis, 1981.

Super G, editor: S.T.A.R.T. instructor's manual, 1984. Available from: Hoag Memorial Hospital Presbyterian, 301 Newport Blvd, Box Y, Newport Beach, Calif 92663.

Trends in Triaging: A Random Survey of Triage Tags Used in the United States, Emerg Prod News 88–91, Oct 1977.

CHAPTER 9

COMMUNICATION WITH THE PUBLIC

Convincing the public to evacuate areas threatened by impending disaster is often difficult. Ben Buerger is shown standing in the ruins of his store on Dauphine Island, Alabama, where he rode out Hurricane Frederic. (Courtesy of Robert Madden, (©) 1980, National Geographic Society.)

In disasters, communication with the public assumes new dimensions not present in routine emergencies. This chapter will address some of the more important points of two aspects: issuing warnings to the public and handling inquiries from the public.

When warning is possible, it may have the greatest potential for saving lives and property, because it allows people to take protective action before impact. The effectiveness of warning requires not only that the message is received, but that it is based on accurate assumptions about human behavior in disasters.

Large numbers of inquiries from the public to emergency and governmental agencies are an almost inevitable consequence of disasters. Often the volume of such information requests can place substantial demands on the recipients. Organized efforts to deal with these inquiries can lessen their disruption.

WARNING

The Value of Warning

Warning can be one of the most important types of disaster communication, allowing the recipients to avoid the threat altogether or to significantly lessen its effects (Mcluckie, 1970:2). A number of disaster countermeasures can be taken as a result of effective forewarning. Probably the most effective is to leave the threatened area before the disaster hits. Other adaptive responses include sandbagging to prevent flooding, boarding up windows to prevent wind damage, mobilizing teams in anticipation of search and rescue activities, or stocking up on food, fuel, water, flashlight batteries, and medical supplies. In a number of disasters, many lives have been saved, even in the face of tremendous property losses, because the affected population received advanced warning (Adams, 1981b:53; Drabek, 1981:87; Quarantelli, 1982c:57).

> **EXAMPLE:** *Tornado, Wichita Falls, Texas, April 10, 1979.* This tornado ranked at 4 on the Fujita scale, placing it in the top 3% of tornado severity. It was one of the widest tornados ever observed and stayed in contact with the ground for a distance of 47 miles, cutting an 11-square-mile path of destruction through the city. It was felt that the injuries and deaths (171 hospitalized and 47 dead) resulting from the storm would have been much greater had there not been an effective warning prior to impact (Quarantelli, 1982a:G54; Adams, 1981b:53; Fox, 1981:7; Glass, 1980:737).

Planning Assumptions

Effective procedures for warning must be based on accurate assumptions about how the public reacts to warning messages. Unfortunately, officials have put

Figure 9-1.   About to be engulfed by a tsunami, a man faces his last moment alone. This wall of water, which reached a height of 55 feet, was generated by an earthquake in the Aleutian Islands. It struck Hilo, Hawaii on April 1, 1946, killing 159 persons. Tsunami warning systems are the most effective means of preventing loss of life in this type of disaster. This photo from the ship *S.S. Brigham Victory*. (Courtesy of Water Resources Center, University of California at Berkeley, Berkeley, California.)

out warning bulletins most cautiously, or withheld warnings until the last minute, because they felt that the inevitable panic would be almost as dangerous as the disaster itself (Dynes, 1981:16; Quarantelli, 1965:107; Quarantelli, 1972:67; Fritz, 1961:664; Drabek, 1986:120).

> **EXAMPLE:** City officials and state police refused to order the evacuation of an eastern resort threatened by an approaching hurricane. They preferred to chance the danger of inaction, because they feared the warnings would result in a panic flight. This was despite urgent recommendations by the Weather Bureau and Coast Guard that the warnings should be issued. It was also despite knowledge that the two routes of escape from the low-lying city would be impassable if the magnitude of the storm was as great as predicted (Quarantelli, 1960:68).

> **EXAMPLES:** Because of similar beliefs about the risk of panic, warnings were played down during the *Rio Grande flood*,

and the *Worcester tornado*. Fear of panic is also why alarm bells were not rung on the collision-doomed ship, *Andrea Doria* (Quarantelli, 1960:68).

The Absence of Panic

However, contrary to popular belief, research has shown that panic is not a common reaction to disasters (Dynes, 1974:71; Dynes, 1981:16,18; Quarantelli, 1960:68; Quarantelli, 1965:107; Quarantelli, 1972:67; Mileti, 1975:57; Drabek, 1986:136; Wenger, 1975:33; Wenger, 1985a:30). This is not to say that panic never occurs, but that it is rare. Furthermore, if it does occur, three conditions appear to be required (Mileti, 1975:58):

—a perception of immediate danger,

—apparently blocked escape routes, and

—a feeling by the victim that he is isolated.

Finally, if panic occurs, it is not widespread or contagious. It is almost always highly localized, with few participants, and of short duration (Quarantelli, 1960:72). The lack of panic in disasters is documented in Table 9-1.

One of the reasons for the belief that panic is common is failure to draw the distinction between evacuation and wild panic. Fleeing a threat is not the same as panic. Sometimes withdrawal is the most intelligent response to a hazard. A panic-stricken individual, however, flees without consideration for others. In contrast, persons who leave an area in an orderly evacuation often assist others to get away (Quarantelli, 1972:68).

> **EXAMPLE:** *Flood, Denver, Colorado, June 16, 1965.* When residents of Denver were threatened by the rapidly rising flood waters, 92% of the families who evacuated left together. This is in contrast to the pattern one would expect from a panic-stricken population (Quarantelli, 1972:68).

Reluctance to Evacuate

Not only is panic flight an uncommon response to disasters, but it is often difficult to get people to leave when disaster threatens (Quarantelli, 1972:67; Quarantelli, 1960:69; Quarantelli, 1965:107; Fritz, 1961:665; Wenger, 1985a:34; Perry, 1985:54) (see Table 9-2).

A similar hesitancy to flee in the face of impending disaster has been documented in dozens of disasters of all sorts. In some of these cases, even the threat of force and coercive measures was not enough to assure evacuation (Quarantelli, 1960:67,69).

Table 9–1.   Absence of Panic in Disasters

Disaster	Observation
Fire Coconut Grove Nightclub, MA 1942	During this fire which has been described as the model of a panic situation, as few as one third of the people may actually have panicked (Keating, 1982:89). Most of the persons did not panic, but calmly gathered their friends and evacuated (Quarantelli, 1972:68).
World War II	In cities such as Hamburg, Hiroshima, and Nagasaki, subjected to mass bombing, panic was not a problem (Fritz, 1961:671).
Atomic bomb explosion Hiroshima, 1945	There was no evidence of mass panic. Many tried to help others (Barton, 1969:148; Fritz, 1961:671).
Tornado White County, AR 1952	People were frightened, but there was little screaming or other uncontrolled behavior. Most did what they could to help themselves and others. Victims were extraordinarily calm and cooperative, even those who suffered from serious wounds or a death in the family (Barton, 1969:1,6).
Tornado Waco, TX 1953	There was little panic. A few survivors wept hysterically, or were dazed, but the largest number immediately joined in spontaneous rescue efforts (Moore, 1958:7).
Coliseum explosion Indianapolis, IN 1963	There was no panic flight or hysterical behavior. In fact, it took a second, less severe explosion to encourage lingering onlookers to evacuate the building (Drabek, 1968:46).
Earthquake Anchorage, AK 1964	There was no panic or hysteria at the hospitals at any time, either in the patients or the staff (Yutzy, 1969:68).
Fire Beverly Hills Supper Club, KY 1977	The National Fire Protection Association report found no evidence of panic related to the 164 deaths (Keating, 1982:89).

cont'd.

Table 9–1.   Absence of Panic in Disasters *cont'd.*

Disaster	Observation
Tornado Grand Island, NE 1980	Among patients in the general and Veterans Administration hospitals, there was concern, even fear on the part of some, but no panic. Behavior generally was marked by cooperation and concern for others (Quarantelli, 1982c:66).
High-rise fire MGM Grand Las Vegas, NV 1980	When the fire department arrived, there was no indication of panic (Best, 1982:21). The behavior of guests was primarily cooperative and altruistic. The only exception was during the first 15 minutes of helicopter evacuation from the roof (Bryan, 1982:46). The 26 persons who died in their rooms showed no signs of panic (Munninger, 1981:40).
Skywalk collapse Hyatt Hotel Kansas City, MO 1981	Hotel employees remained calm, helping people and directing others (Dunlap, 1981:7).
Metrorail subway crash Washington, DC 1982	The wrecked passenger cars were evacuated in an orderly fashion; there was no panic (Adams, 1982:54).

PRINCIPLE

Panic is not a common problem in disasters; getting people to evacuate is.

Premature Return of Evacuees

Even if they have been convinced to evacuate, inhabitants may return while the threat is still present.

> **EXAMPLE:** Four months after the initial war-time evacuation of British cities, over 60% of the population had returned. This was in spite of official warnings that these areas were prime targets for air raids and rocket attacks. Although the

Table 9–2.   Reluctance to Evacuate in Disasters

Disaster	Observation
Bombing of Britain WW II	Only 37% of the mothers and children evacuated London during the raids. The town of Bootle was bombed every night for a week, and 60% of its houses were hit at least twice each, and only 10% escaped serious damage. Nonetheless, 25% of the town's inhabitants remained to sleep in their homes during the raids (Quarantelli, 1960:69).
Hurricane Florence Florida 1953	Despite intensive warnings, 66% of the residents in a town in the hurricane's path refused to leave their homes (Quarantelli, 1960:70).
Hurricane Carla Texas 1961	About 70 to 80% of Galveston residents stayed on the island even though most knew that they would eventually be cut off from the mainland (Davenport, 1978:19).
Hurricane Frederic Mississippi 1979	As the winds intensified, in spite of a massive evacuation effort, nearly half the residents refused to leave their homes (Drabek, 1981:141).

cities were being bombed nightly, even children were being brought back to London in large numbers. A similar pattern occurred in Germany in spite of governmental efforts to discourage it by withholding ration cards and schooling (Quarantelli, 1960:70).

EXAMPLE: *The Mt. St. Helens, Washington, Volcano Eruption, May 18, 1980.* Risking a $500 fine and 6 months in jail, many residents circumvented the barricades around the threatening volcano. Taking to the back roads, they went to check on their property and retrieve belongings (Kilijanek, 1981:57).

Reasons for Hesitancy to Evacuate

There are a number of reasons why persons hesitate to evacuate in the face of threatening disaster. They may not be convinced that they are actually at risk;

they may wish to stay and protect their property; or they may want to assure the safety of other family members before leaving.

Perception of risk.

The most common reason people do not evacuate is that they do not believe they are in immediate danger (Perry, 1985:53). People tend to interpret observations in light of what they expect to happen. Since disasters are such a rare experience for most people, the natural reaction to warning is disbelief. This effect is magnified when the warning is related to a type or severity of threat that is unlikely to occur in the recipient's area. Thus, residents of Kansas are likely to heed a springtime tornado warning. On the other hand, a flash flood warning to those living near a quiet stream which has never flooded before is less likely to be taken seriously (Drabek, 1985b:12).

In addition, the thought of impending disaster is one that most people would prefer to avoid, and thus they may tend to deny it. This is not to say that people will necessarily ignore warnings, but if there is any ambiguity or contradiction in the warning information, it is often interpreted as evidence that the best rather than the worst situation exists. For example, an air raid siren may be taken for another test, or a mistake. Often, only after these possibilities are shown to be untenable are other less pleasant interpretations considered (Mcluckie, 1970:40; Drabek, 1986:73,82).

In most cases, the first reaction to a warning, if it has not been expected, is to try and confirm its validity. One way this is carried out is by observing the behavior of others. The failure to see them behaving in an alarmed manner may lead to a discounting of the warning as a mistake, misunderstanding, or overreaction. Attempts to validate the information may take the form of phone calls to relatives, friends, or public safety agencies. Another common response is to turn on the radio or TV for further information. Validation can also take the form of assessing the warning in the context of environmental cues (Mcluckie, 1970:41; Drabek, 1985b:12; Drabek, 1986:83,113,123; Perry, 1985:67,79).

> **EXAMPLE:** *Tornado, Grand Island, Nebraska, June 3, 1980.* The warning sirens were heard frequently from April through late summer, but the last actual tornado was in 1857. Nevertheless, on the evening of June 3, they were not heard with the usual complacency, because the skies began to look uniquely ominous. Many people were acutely aware of the weather, turning on the radio to get further information, even before the sirens sounded. When they did go off, furthermore, they were heeded. Thus, in spite of bearing the full brunt of six twisters that flattened one fifth of the town, the town lost only five of its 40,000 residents to the storm (Quarantelli, 1982c:57).

Even when persons hear a warning and accept the fact that a disaster is threatening, they may still fail to evacuate because they don't believe they are in personal danger (Perry, 1985:70). This may be because the warning does not carefully specify the severity of the forces involved, or because previous disasters have failed to materialize after warnings or have been of less magnitude than predicted.

> **EXAMPLE:** *Hurricane Audrey, Lower Cameron Parish, Louisiana, June 27, 1957.* A large number of the 400 deaths from this storm were from an area where the residents thought the rising waters would not reach the ridges on which they lived. The warning messages issued failed to make it clear that they would (Mcluckie, 1970:7; Bates, 1963:13).

Protection of property.   Another common reason persons hesitate to evacuate is because they want to protect their property. In some cases, this has been because of the fear of looting, but many times, it has been due to

Figure 9-2.   As in this wildland fire, residents are often hesitant to evacuate in disasters, preferring to remain and protect their property. In such situations, the absence of effective and convincing warning procedures can lead to loss of lives. This photograph of "The Fourty-Niner Fire," of Nevada County, California in September, 1988, is a good example of need for warning procedures. (Courtesy of *The Union*, Grass Valley, California.)

the desire to protect property against the environmental threat (Perry, 1985:53, 153).

Safety of family members. Persons in disaster-threatened areas often hesitate to evacuate until they have assured the safety of other family members. Often, this means that the family will evacuate as a unit, but only after all family members are located or accounted for. Sometimes this definition of "family" even applies to household pets, particularly dogs (Perry, 1985:60,72; Drabek, 1986:84,114,116).

Improving the Response to Warning

In contrast to the belief that people will flee in panic when warned of a disaster threat, the chief difficulty is in getting people to evacuate. There are several factors which can enhance warning effectiveness.

Context of the Warning Message

The credibility of warning is enhanced it if is issued in a context consistent with a condition of urgency. For example, if a TV station issues a tornado warning and then suspends regular broadcasting to follow the storm's progress, the viewer's perception of urgency is enhanced. If, however, the warning is followed by a return to normal programming, the threat is not taken so seriously. In some cases, information given out with the warning can have a neutralizing effect on it (Mcluckie, 1970:33; Drabek, 1985b:13; Perry, 1985:44,58).

> **EXAMPLE:** *Rio Grande Flood, Piedras Negras, Mexico, June 27-30, 1954.* Two loudspeaker cars were "drafted" from a local theater to assist in warning the public. Reportedly, one of them issued the following alert, "An all-time record flood is going to inundate the city. You must evacuate immediately. (Pause) The _____ theater is presenting two exciting features tonight. Be sure to see these pictures at the _____ theater tonight." (Mcluckie, 1970:33)

The validity of *past* warning messages can also influence believability. Persons living in areas frequently warned about approaching threats, but which rarely sustain a severe impact, tend to discount the seriousness of subsequent warning messages (Mcluckie, 1970:23,26,33,37; Drabek, 1986:77,93).

Consistency and Repetition

Hearing repeated warnings increases the likelihood of taking protective action (Drabek, 1986:61,76; Drabek, 1985b:13; Adams, 1981b:15,53). Consistency of the warning information from different sources enhances its effect. When different sources of information convey similar information about the threat, persons trying to confirm the warning are more likely to heed it (Drabek, 1981:76,95,113). The greater the number of different warning sources, the larger the number of people contacted (Moore, 1958:212; Adams, 1981b:15,53,57; Fox, 1981:8; Glass, 1980:737; Perry, 1985:41).

Legitimacy of the Source

Warning messages are more likely to be believed if they are issued by official governmental authorities, such as the police, civil defense, fire department, the governor, or the mayor's office (Drabek, 1986:75,104).

Specificity

The specificity of warning influences its effectiveness. Recipients of warning information need to know more than just the fact that there is a disaster threat. They need information that indicates how the threat will affect them personally. Although sirens can alert a large number of people, they carry the least specific type of information. They do, however, get people to turn to potentially more specific sources of information, such as the mass media, especially if sirens go off in inclement weather (Mcluckie, 1970:31; Drabek, 1986:92; Quarantelli, 1982c:57,61). Helpful information is that which states *in terms clear to the recipient*, the urgency of the situation, the likelihood of impact, and the specific localities at risk. "Terms clear to the recipient" implies the need for foreign languages in certain ethnic communities. It also infers that the terminology used is meaningful to its audience. For example, saying that flood waters will crest 5 feet above flood stage may convey less meaning than saying that the waters will cover the courthouse stairs. Even in parts of the country where tornadoes are common, terms like "tornado watch" and "tornado warning" are misunderstood by over a third of the public (Drabek, 1986:74,106,335; Perry, 1985:67,71; FEMA, A-50; Kreimer, 1980:21).

Information on Courses of Action

For a warning to be effective, it has to do more than just alert the public of a disaster threat. Information has to be given regarding appropriate protective actions that should be taken. Protective actions may be obvious to some living in disaster prone areas, but to many, especially in technological accidents, the proper courses of action may be less obvious (Perry, 1985:66,79).

> **EXAMPLE:** *Nuclear Power Plant Accident, Three Mile Island, Pennsylvania, March 28, 1979.* Of those living within 15 miles of the reactor, 39% (144,000) evacuated. Of those that failed to

evacuate, 62% indicated it was because they were not instructed to do so (Perry, 1985:53).

Invitations from Relatives

Those living in disaster-threatened areas are more likely to evacuate if they are
encouraged by invitations from relatives and friends outside the impact area
(Drabek, 1985b:17). This is because people prefer to seek shelter with friends or
relatives rather than at public shelters. It also reflects the effects such invitations
have in confirming the danger. Encouraging this activity can enhance compliance with evacuation advisories (Drabek, 1986:81,86,118).

TelePatrol International

One of the most potentially effective technological approaches to warning is the
use of computers to contact selected populations by phone and give recorded
warning information. This technology is now widely available through "Tele-
Patrol International." TelePatrol International is a non-profit organization
formed for the purpose of bringing this alerting system into widespread use.

Using TelePatrol, local communities can list all residential and business
phone numbers by category in a computer data bank. The system can then be
programmed to rapidly dial every number and announce a recorded message
or warning. If the number is busy, the system will redial until it gets a response.
This allows large numbers of people to be warned of a hazard, even if they have
their radio or TV turned off, or if they are asleep.

The system allows listing the phone numbers according to geographic location. Thus, if a tanker truck carrying hazardous material is involved in an
accident, all those living in the immediate vicinity can be warned to evacuate.
Phone numbers can be categorized by features other than location. For example, non-English speaking households can be identified and the message given
in the appropriate language. Households with deaf residents or those without
phone can also be indicated. Furthermore, the system will identify those
addresses that have failed to respond. The system can be used to recall public
safety or hospital personnel. It can also be used to call sources of special
supplies or equipment needed for a disaster. In addition to its ability to carry
messages to the public, TelePatrol can also receive and collate information. For
example, it can ask those who are disabled and need transportation assistance
to indicate that need.

In addition to uses in disaster warning, TelePatrol has a number of uses in
daily emergencies and law enforcement. For example, it can quickly give a
neighborhood the description of a child who is lost in that area. It can also
expand the distribution of the message as time passes and the area of search
increases. If a report is received that a food or pharmaceutical product has been

tampered with, every grocery store and drug store in the area can be rapidly notified. The possibilities are endless.

A participating community has to pay a $1 fee and agree to assist in establishing a local TelePatrol Board of Directors from its business and community leaders which would be one of the outreach mechanisms for raising funds for the local program. The community is also required to appoint one liaison from existing government personnel to coordinate the public relations and political aspects of the program and another to coordinate operations and technical aspects. TelePatrol International is responsible for fund raising with the assistance of the liaisons and the support of the local government chief executives. All of this funding stays in the local community's TelePatrol program. For more information, see Additional Reading at the end of the chapter.

Coordination of the Warning Process

The process of warning is complicated by the fact that it requires the accomplishment of a number of tasks, and because these may have to be carried out by different organizations, coordination is required among them. For example, the determination that adverse weather will lead to unusually heavy rainfall is usually made by the Weather Bureau. This might alert the local flood control authorities to the possibility of flooding and the subsequent detection of impending dam or levee failure. The decision to issue an evacuation directive might then come from the sheriff's department or the office of the county executive. But the conveyance of the message to the public is often carried out by local commercial radio or TV stations (Stallings, 1971:34; Dynes, 1981:9).

INQUIRIES FROM THE PUBLIC

Many inquiries from the public are an almost inevitable consequence of disasters. Often the volume of such information requests can place substantial demands on mayor's offices, police and fire departments, hospitals, news agencies, and other sources of disaster information. The bulk of these inquiries are of three types: inquiries to confirm the validity of warnings; inquiries about the welfare and location of missing loved ones; and instruction on what to do (Scholl, 1984:289; Quarantelli, 1965:110; Fritz, 1956:14, 38; Lantis, 1984:52; Ross, 1982:64; Drabek, 1986:85).

These inquiries can be disruptive for several reasons. Their sheer numbers can jam local telephone circuits; essential public safety activities are interrupted as agency personnel attempt to answer the inquiries; and it may be very difficult to collect and collate the information necessary to respond (Seismic Safety Comm, 1983:16,33; Drabek, 1986:85; Worth, 1977:160; Moore, 1958:8).

Efforts to Locate Loved Ones

Because the United States is a very mobile country, family members and loved ones are often separated from one another. Nearly every family in the country has blood relatives living in other parts of the nation or in foreign countries. Families living together are usually temporarily separated at different times of the day (Fritz, 1956:36). This fact is of special importance when disaster strikes, because there is an immediate and frantic effort of loved ones to locate those thought to be disaster victims.

The mass communications media not only quickly notifies the world of these events, but many times their versions are greatly dramatized, if not distorted. In addition, news reports usually do not give specific information about the exact location of a disaster, or details to indicate who has or has not been involved. Even in disasters with only a few hundred homeless, injured, or killed, the total number of personal welfare inquiries may be in the tens and hundreds of thousands (Fritz, 1956:22,36,37,63). Few public safety agencies, hospitals, emergency organizations, or governmental bodies are prepared for the deluge of inquiries after a disaster, and the results can be literally paralyzing.

Jammed Telephone Circuits

When people hear of a disaster that might involve loved ones, the first response is to telephone them. If loved ones cannot be located where expected, information is sought by phoning hospitals, police departments, fire departments, relief agencies, newspaper offices, or the city hall. Since it only takes a small percentage of the population using the phone simultaneously to overload the system, phone lines in the disaster area quickly become jammed (Drabek, 1985b:13; Stallings, 1971:34).

> **EXAMPLE:** *Tornado, Waco, Texas, March 21, 1952.* After the tornado struck, incoming calls were so numerous that outgoing calls were delayed for as much as 6 hours. Similarly, the communications resources of the telegraph office, the post office, "ham" radio operators, and the MARS (Military Affiliate Radio System) were inundated (Fritz, 1956:19).

> **EXAMPLE:** *Flash Flood, Big Thompson Canyon, Colorado, July 31, 1976.* As word of the disaster spread across the country, concerned relatives deluged the Denver office of Associated Press, often blocking incoming reports from reporters at the scene (Ritz, 1980:195).

Community organizations considered sources of disaster information (police departments, hospitals, municipal offices) have their switchboards so flooded with calls that all communication in or out of the organization is prevented. Therefore, emergency procedures dependent on telephone communication cannot be carried out.

Traffic Congestion in the Disaster Area

When loved ones cannot find the information they seek by phone, those within traveling distance try to seek it out in person.

> **EXAMPLE:** *Gas Main Explosions, Brighton, New York, September 21, 1951.* When a city gas main pressure reducing valve failed, pilot lights of some gas appliances were extinguished by the gas pressure surge. The houses then filled up with gas which then exploded when it reached the pilot light of another appliance or when someone turned on an electrical switch. Forty-one houses were damaged or destroyed and twenty-seven casualties resulted—three of them fatal. Getting in contact with all immediate family members was one of the first responses. At the time of the disaster, most of the men were away working in Rochester. When they heard of the explosions, they tried to call home. Most of them, unable to get through the jammed phone exchange, tried to drive back to Brighton. They soon found themselves tied up in the traffic jam that developed. In many cases, even when the men eventually reached the edge of Brighton, they were stopped by road blocks. Despite all this, a number of them took to the back roads and were able to work their way into the disaster area (Fritz, 1956:11,13,15,16; Marks, 1954:Appendix B-2:36).

> **EXAMPLE:** *Tornado, Worcester, Massachusetts, June 9, 1953.* Twenty minutes after a tornado swept through the city, the traffic jam on the roads leading to the area was formidable. It interfered with the passage of fire, police, ambulance, and other emergency vehicles. Complicating the traffic problems were "the hundreds of fathers, mothers, sons, and daughters of the disaster area residents." They were abandoning their cars and running into the impact area to find and help their families (Wallace, 1956:74).

> **EXAMPLE:** *Atomic bomb explosion, Hiroshima, Japan, August 6, 1945.* The city was evacuated after the explosion. But, less than 24 hours after the evacuation, "thousands of refugees came streaming back into the city . . . road blocks had to be set up along all routes into the city because there were so many

people who wanted to search for missing relatives or inspect the damage . . . " (Fritz, 1956:12).

Similar examples can be documented from many other disasters. Hospitals, police departments, and other places of disaster activity are also swamped with people who go there trying to locate loved ones (Fritz, 1956:37).

Advice

Adding to the disruptive effects of those seeking information about loved ones, are those seeking advice about disaster-related problems. Questions may be asked about such things as whether or not the callers' homes are in an area threatened by flooding; whether or not food exposed to the chemical cloud can be eaten safely; and where the caller should go to donate blood. They often present organizations with requests for information which the organizations may not be prepared to provide and which require the diversion of resources to obtain (Quarantelli, 1965:110; Fritz, 1956:15; Raker, 1956:44; Drabek, 1968:49).

Management of Inquiries: Disaster Public Information Centers

Answering these questions is often impossible unless there exists a system for the various organizations to share information (see Fig. 9-3). For example, inquiries about missing loved ones may require the collection of information from the coroner, law enforcement agencies, public shelters, and local hospitals (Quarantelli, 1965:110; Quarantelli, 1983:83; Scholl, 1984:289; Yutzy, 1969:121; Lantis, 1984:52). This requires the establishment of a central, multi-organizational public information office for the disaster. The responsibilities of this office are three-fold: 1) information for the disaster community (warning and instructions); 2) information for the press and dignitaries (discussed in Chapter 10); and 3) information for outsiders inquiring about loved ones.

Establishing a regional system of disaster public information centers can help in dealing with outside requests about loved ones. This system, composed of local information centers connected with regional centers, channels as many outside welfare inquiries as possible *away* from the centers of emergency activity.

Information on dead, relocated, and injured casualties is collected from designated officials at local hospitals, police departments, morgues, coroners offices, Red Cross, Salvation Army, and community shelters. It is then transmitted to regional centers in other parts of the state or country. Pre-arranged agreements with the media provide the public with toll-free numbers to these regional information centers where they might learn if their relatives are listed as known fatalities, or if relocated or injured, where they are now.

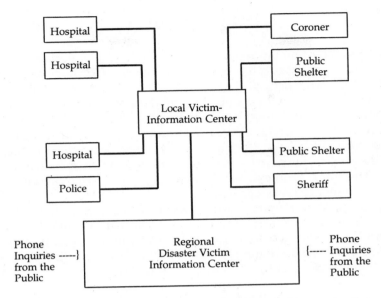

Figure 9-3.   The use of regional disaster information centers to divert public inquiries from emergency and public safety agencies.

The local information center in the disaster-stricken community also collects information that would allow others to learn if their loved ones were victims of the event. For example, information on the scope and severity of the event allows callers to determine if a loved one's residence was in the seriously impacted area. If the event is an airline crash, the information includes the points and times of departure and arrival, the airline company and flight number, the passenger list, and the streets and block numbers in the crash impact site. This information is also provided to the regional centers and to the press. The provision of this detailed information in news accounts of the event helps to decrease the number of viewers who feel they may have a loved one in the disaster area.

> **PRINCIPLE**
> Inquires about loved ones thought to be in the impact zone are not likely to be discouraged, but can be reduced or channeled in less disruptive ways, if the needed information is provided at a location away from the disaster area.

SUMMARY

Disaster-stricken communities often have difficulties communicating with the public. Issuing warnings is one of the most important methods of averting the destructive consequences of disasters. In some cases, an effective warning process may depend on the cooperative interactions of multiple organizations: those who detect the disaster threat, those who decide that a warning should be issued, and those who convey the warning to the public. The public's response to warning is not a simple stimulus-response reaction. Rather, members of the public often have to be *convinced* that they are in immediate, personal danger. The source, context, and repetition of the message can influence the warning's influence on public behavior.

In disasters, news reports generate worldwide concern in those who think they may have loved ones in the impact area. The usual result is that organizations in the disaster area are inundated with inquiries about these persons. Although these welfare inquiries cannot be stopped, effective planning can reduce them or channel them so they are less disruptive.

PLANNING CHECKPOINTS

[ ] Do those responsible for issuing warnings to the public understand that widespread panic is not a common problem in disasters, but that convincing people to evacuate is?

[ ] Does your community's disaster planning and training address which organizations and persons are responsible for the various aspects of warning: detection of the threat, decision to warn, and dissemination of warning?

[ ] Does your warning process take into consideration the importance of the warning context? The legitimacy of the warning source? The importance of warning repetition and consistent, multiple warnings from different sources?

[ ] Does your planning include provisions to disseminate information that will help members of the public determine that they don't have loved ones impacted by the disaster (such as information accurately describing the geographical boundaries of the disaster, the involved aircraft destination, flight number, and list of uninjured passengers)?

[ ] Do you have a regional system for collecting disaster victim information and providing it to the public at a site away from the disaster response activity?

ADDITIONAL READING

Mcluckie B, The warning system in disaster situations: A selective analysis, Report Series 9, 1970. Available from: Disaster Research Center, University of Delaware, Newark, Del 19716.

Perry RW, Comprehensive Emergency Management: Evacuating Threatened Populations, 1985. Available from: JAI Press, Inc, 36 Sherwood Place, Greenwich, Conn 06830.

Quarantelli EL, Evacuation Behavior and Problems: Findings and Implications from the Research Literature, Book and Monograph Series 16, 1984. Available from: Disaster Research Center, University of Delaware, Newark, Del 19716.

TelePatrol International, TelePatrol Management, Inc, P O Box D, Sunny Lane, Beach Lake, Penn. 18405, (800) 255-4583.

THE MEDIA: FRIEND AND FOE

The magnitude of media response to future disasters it likely to be much greater than that shown in this photograph of the 1971 San Fernando Valley, California earthquake. (Courtesy of Los Angeles County Fire Department, Los Angeles, California.)

Many emergency managers have been frustrated when they have had to divert much needed time and resources to address the demands of the media, while at the same time trying to mount a multi-organizational disaster response under condi-

215

tions of extreme urgency and uncertainty. Well-planned interactions with the media, though, can be of critical importance in
decreasing the loss of life and property. In those types of
disasters where warning is possible, accurate, timely, and
consistent information conveyed by the media can be one of the
greatest factors preventing death and injury. It has been suggested that difficulties with the media occur because emergency agencies do not understand how the media operate in
disasters and how to deal with them in an effective way. This
chapter identifies some of the important patterns of media
behavior in disasters and suggests ways of optimizing the
interaction between the media, the public, and disaster response agencies.

DISASTERS ARE A MEDIA EVENT

Disasters are a significant source of news. In fact, one study by Gans estimated
that 25% of all news stories involve natural disasters, technological hazards, or
civil disturbances. Why is this the case? One reason that has been suggested is
that mass media news reporting is an entertainment business. Often, news
reports are called "stories," and reporters are encouraged to find events that
have the same attributes as that of good fiction: drama, conflict, problems,
solutions, and rising and falling action. Disasters offer all of these characteristics, and for television they present the additional advantage of great attention-
grabbing visuals (Wenger, 1985b:2; Larson, 1980:79,119).

> "From the journalistic point of view, a natural disaster has all
> the ingredients for the 'perfect media event' (especially for the
> electronic media). It's brief, spectacular, often mysterious, ac-
> tion-oriented, and portrays human suffering and courage."
> (Bolduc, 1987)

Another reason that has been given to explain why disasters attract so much
media attention is that they are generally so easy to cover.

> "Television likes to cover disasters. All it takes is a film crew or
> two, a public shelter where victims can be photographed, a few
> shots of rising water, destroyed homes or trailers, some high
> turf, and an authority figure to interview and you have one
> minute and thirty seconds of dynamite, visual coverage."
> (Wenger,1985b:3)

Evening news broadcasts are highly profitable programs for most local stations. They are also valuable for local network affiliates because they attract an

audience that also tends to watch the network's national news broadcasts. Therefore, in all major markets there is intense competition for local news audiences, and a considerable amount of "show business" and entertainment has been injected into programming to help attract these audiences (Larson,1980:79).

THE MEDIA AS "FRIEND"

Public officials sometimes decry the mass media. Many feel that they would be able to carry out their disaster duties in a calmer atmosphere if the media were not there to play "sideline quarterback," criticizing actions and questioning decisions. However, in spite of this feeling, the absence of the media in disasters can create enormous difficulties (Scanlon, 1985:123).

The mass media, and the electronic media (TV and commercial radio) in particular, are the most imporant source from which the public obtains information on disasters (Wenger, 1980:241; Wenger, 1985a:62). The importance of the private sector news media as a communications system in the United States is reflected in the following quotes:

> "The entire governmental structure of the United States—Department of Defense, FEMA, NASA, all of the state and local governments—together have sophisticated communications systems costing dozens of billions of dollars; they are the envy of the rest of the world. But in our society that marvel of electronic wizardry is incapable of crossing the threshold of the American home, or entering the confines of the American automobile to communicate directly to the citizen. For that link to the public the emergency management community is totally dependent on the private sector, the news media."

> "The development of a reasonable, effective and constructive working relationship between the emergency manager and the media in disasters should be a high priority goal of the emergency management community. The fundamental responsibility of all governments is assuring the safety and well-being of its citizens. That mission cannot be carried out in an information vacuum. The citizen must know if and when he is in danger, and if and when the danger has passed. And he has a right to know about the fate of his neighbors." (Holton, 1985:6)

The mass media can play a number of important roles that help to lessen the effects of disaster (Comm on Disasters, 1980:vii). They transmit warnings of impending tornadoes, hurricanes, floods, tidal waves, and volcano eruptions.

Commercial radio and television are the most frequent sources from which the public receives initial warning about an impending disaster. Even when initial warning is from another source such as sirens, people turn to the media for further information (Drabek, 1986:91,113; Quarantelli, 1982c:61; Moore, 1958:212; Kreps, 1980:61). Indeed, people often react to warning sirens with disbelief until such confirmation is obtained (Kreps, 1980:61). When media warning messages are clearly worded, authoritative, and consistent, they can be very effective in stimulating appropriate protective activity (Wenger, 1985b:18).

On occasion, the media have even been known to initiate such warnings without awaiting official requests:

> **EXAMPLE:** *Tornado, Topeka, Kansas, June 8, 1966.* A local radio station had its own alert plan for severe weather situations. This involved the dispersal of mobile radio units that would make on-the-spot reports if a funnel cloud was sighted. The radio station's own warning broadcast was not only heard before the sounding of the public alert system, but also 30 minutes before the Weather Bureau's official teletype warning (Kreps, 1980:60).

Often, it is not only the public that receives useful warning and information from the media. When, as often occurs, there is inadequate communication among emergency response organizations, their best source of information may be from the mass media (Wenger, 1985b:18).

> **EXAMPLE:** *Flash Floods, Terrace, British Columbia, October 30, 1978.* CFTK, the local radio and TV station, was often the first source of information for the public about disaster-related problems. "Individuals in trouble called the . . . station rather than an official agency. The station then quite quickly passed that information on to the various authorities. The station also was the first place to define the extent of the emergency situation. A number of local officials said that they had not viewed the situation as being close to a disaster until they heard the reports of the media. The various disaster response agencies tended to work on their own and not share information. Thus they learned the overall situation only by listening to radio and/or watching television (Scanlon, 1980:260).

> **EXAMPLE:** *Metro Subway Crash, Washington, DC, January 13, 1982.* Little useful information was received from the accident site, and information from official governmental sources was conflicting. Most of the hospitals learned of the crash either

from incoming emergency medical services personnel or from the news media (Edelstein, 1982:161).

In a number of disasters, it has been observed that the media has conveyed important information about the disaster useful to government officials and relief agencies (Moore, 1958:189; Wenger, 1985b:18; Edelstein, 1982:161; Killian, 1953:S-2).

> **EXAMPLE:** In a forest fire disaster one radio station "became an emergency appendage for a number of disaster-related organizations, including the Civil Defense, the police and fire departments, the county sheriff's office, and the Salvation Army. The station not only served as an interagency communications link, but it also collected information about the fire for emergency organizations and sent that information to them through its mobile unit 'live coverage' of the disaster." (Kreps, 1980:63)

The mass media may perform a number of other useful functions to aid disaster-relevant organizations and the public:

- They may convey instructions to the public as to how they can lessen or deal with the effects of the disaster. They may help to educate the public about how to prepare for disasters (Kreps, 1980:59; Wenger, 1985b:17).

- They may stimulate donations from other parts of the country which (if handled properly so as not to overwhelm the stricken areas with unneeded supplies) can help speed recovery (Kreimer, 1980:18; Comm on Disasters, 1980:4).

- They may draw attention to natural and technological hazards and stimulate public support for actions to prevent or mitigate disasters. They may help to overcome public and governmental apathy by drawing attention to disaster risks and inadequate preparedness (Stevenson, 1981:36; Drabek, 1986:62).

- They can help to minimize the number of inquiries from anxious loved ones by providing accurate information about the severity and scope of the disaster and by publicizing lists of survivors (Kreimer, 1980:18; Kreps, 1980:46).

- When other means are not available they can be used for point-to-point and person-to-person communications if necessary to expedite rescue and recovery (Yutzy, 1969:103,122).

- They will often withhold news if they consider it to be dangerous to release to the public (Scanlon, 1982:18).

- Good publicity from media coverage is a factor that may facilitate future funding, donations, or re-election for those involved in disaster response or management (Larson, 1980:89).

THE MEDIA AS "FOE"

Emergency managers have often expressed frustration and consternation with media operations in disasters (Wenger, 1985b:5; 1982). They feel that the media complicate the tasks at hand and that paying attention to media demands diverts their attention from urgent matters like casualty care, search and rescue, and evacuation. As those responding to the Mt. St. Helens eruption discovered, trying to organize all of their forces into a unified, multi-organizational effort is a monumental task. This is even more painful when it has to be carried out under the scrutinizing eyes of the international press (Kilijanek,1981:72).

Demands on Resources, Facilities, and Officials

The media will make demands on communications, transportation, and other local resources. In situations where the disaster has reduced the available functioning communication and transportation systems, these demands are even more noticeable. In extreme cases, their demands may completely tie up any surviving transportation and communication facilities, and local officials may find themselves responding more to the needs of the media than to the disaster situation. The media have other needs too, and expect local emergency personnel to fulfill them. These needs include adequate lighting, electrical power, work space, and the provision of officials to give them information or take them on tours of the disaster area (Wenger, 1985b:7; Scanlon, 1982:17; Scanlon, 1985:124; Rosow, 1977:87).

Reporters have been described as descending on disaster officials like "wolf packs" and overwhelming them with demands for specific information the type of which is seldom available in the early phases of the incident. They will often pressure officials for "exact figures" on the number of deaths and injuries, the amount of property damage, and the quality of aid expected. In technological disasters they will ask why and how such an accident could happen. If officials say they don't know the answers, questions may be raised as to their competence. In many cases, officials are pressured to offer estimates. Unfortunately, estimates obtained from various sources may prove to be conflicting or inconsistent, which also paints a picture of organizational incompetence (Wenger, 1985b:6; Scanlon, 1985:124; Seismic Safety Comm, 1983:70; Rosow, 1977:86).

Distortion of Facts

The entertainment aspect of news broadcasts focuses on the dramatic and unique aspects of disasters, favoring the reporting of the unusual versus the typical or representative events (Quarantelli, 1981b:63; Larson, 1980:93). It has been argued that this perspective perpetuates common myths about disasters. It also leads to an exaggerated picture of the disaster's magnitude, a factor which tends to magnify the problems of overresponse and convergence.

> "Media reports frequently failed to include qualifiers about uncertain information . . . (they) do not seem to want to accept the reality that disasters are full of ambiguity about precisely what happened. The media persist in trying to be precise, and in the process, blunder into inaccuracy." (Kreps, 1980:66)

Perpetuation of Disaster Myths

Decades of research have revealed that a number of popular beliefs about what happens in disasters are incorrect. It has been argued that the persistence of these "disaster myths" is at least partly attributable to the images portrayed by news reports (Bolduc, 1987).

For example, documented cases of panic flight in disasters are extremely rare. In fact, a much more significant problem is getting people to leave their homes, even in the face of certain danger. Nevertheless, the news media seem so preoccupied with panic that the term is loosely applied to describe even orderly evacuations.

Another example is looting. Although documented cases of looting are rare in domestic disasters not involving civil unrest, rumors of looting are commonly reported by the press. Even when it has been correctly observed by the press that looting did not ensue, this has often been couched in terms to suggest that this was the exception rather than the rule. What is reported is that the National Guard or police have instituted measures to prevent looting. However, what is not reported is that no looting, in fact, occurred (Wenger, 1985b:10; Tierney, 1985b:31; Drabek, 1985b:18,21).

Sometimes "what should happen" in disasters has been so well ingrained, that if the story did not conform to the expectation, the reporter shaped it so it did.

> **EXAMPLE:** *Nuclear Reactor Accident, Three Mile Island, Pennsylvania, March 28, 1979.* Media crews advised people to get off the streets because they felt deserted streets were appropriate for such a situation (Scanlon, 1985:124).

Exaggeration of Disaster Impact

Another contention of some disaster researchers is that the media's preoccupation with the dramatic accentuates and exaggerates the destructive magnitude of disasters. This has been labeled the "Dresden syndrome" (the media make every tornado-stricken community look like Dresden after it was bombed in World War II). News films and photographs focus on scenes of destruction, but not upon the surrounding undamaged areas. The audience is often lead to believe that the whole community lies in ruins on the basis of intense coverage of damage which may, in reality, be limited to a few buildings or blocks.

> **EXAMPLE:** *Earthquake, Anchorage, Alaska, March 27, 1964.* One of the widely circulated news photographs of the earthquake showed a wrecked department store. The damage that was the focal point of the picture was impressive, but the buildings across the street were undamaged, with even their large plate-glass windows still intact (Quarantelli,1972; Walker,1982:24).

The impression of widespread destruction is also left by the human interest story of the family that has suffered great loss from the disaster. When these news stories are told, it is not always reported to what extent they represent the disaster's impact. The viewer is often left with the impression that the family's plight is the typical situation of the residents in the disaster area, even when this is not the case (Wenger, 1985b:13; Quarantelli, 1972).

It should be said at this point, however, that the inaccuracy of news reports cannot be attributed to the media alone. There are questions about the extent to which the media introduce distortion and to what extent they are merely passive disseminators of inaccurate information provided by official sources (Kreps, 1980:44; Hartsough, 1985:282).

Interference with Disaster Operations

Effects on Over-response and Inquiries

Exaggerated media coverage has been suggested as contributing to the inundation of inquiries by anxious loved ones. It has also been blamed for the over-response of resources that typify disasters, complicating their management (Wenger, 1985b:13; Quarantelli, 1983:68; FEMA, 1984a:85; Scanlon, 1985:124).

Decision-making

The media have been called a source of interference in local decision-making, pressuring officials to indicate what action will be taken before the officials are ready to decide what to do. Sometimes the media even report greater precision about the intended actions than actually exist. This factor leads to credibility problems for the officials involved (Kreps, 1980:68).

"The analysis process itself requires breaking issues down into components, studying them, and manipulating the possible alternatives. That takes time. When an emergency is imminent or in progress, there is little or no time. Adding to that are the demands of media representatives who are trying to meet their own deadlines. Radio and television demand immediate response; the greater the emergency, the greater the urgency in putting something on the air. The possibilities of error are magnified. . . . While leaders are attempting to gather facts and make assessments, the press is pushing. To news people pursuing a disaster, there is not time; information is demanded immediately and, if officials cannot provide it, then it will be gathered from unofficial sources. Another element of traditional policy analysis is muted or removed entirely. Therefore, the structure of the analytical process is changed. Steps are omitted. Decisions are made with partial facts and without full appreciation of the ramifications. Further, because of the media-imposed time factors, a decision once made and announced is very difficult to reverse." (FEMA, 1984a:85)

Mass media activities may even subvert the adherence to disaster plans and alter the disaster response decision-making process.

EXAMPLE: *Nuclear Reactor Accident, Three Mile Island, Pennsylvania, March 28, 1979.* "The Commonwealth of Pennsylvania had a proposed emergency response system for an accident at fixed-site nuclear facilities prior to the accident at Three Mile Island. Basically, the plan centered around the major coordinating role that was to be performed by the Pennsylvania Emergency Management Agency. However, by the later stages of the disaster, the plan had been abandoned. . . . The entire system had evolved into that of 'emergency management by press conference.' Under the stress of monumental media attention and demands, state and federal authorities centralized all decisions and information-distribution within the Governor's office. This alteration effectively isolated the state's emergency management system not only from active involvement in decision-making, but also from the receipt of information. Local and state emergency management officials, who had planned to be centrally involved in the response, often

found it necessary to listen to radio and television press conferences in order to find out what was happening." (Wenger,
1985b:8)

The presence of the media can limit the alternatives available for response
and make alternative contingency planning difficult. For example, the local
mayor, considering the possibility of violence during a labor dispute, might
hesitate to ask the governor to put the National Guard on standby. If word leaks
out to the press, the situation could worsen. In a toxic spill disaster, if the media
learns that mortuaries are being put on alert, this might conflict with the
officials' simultaneous desire to calm the public (FEMA, 1984a:82).

Rescue and Recovery

The convergence of media personnel at the disaster site has been reported to
physically interfere with response and recovery operations.

> **EXAMPLE:** *Earthquake, Coalinga, California, May 2, 1983.* The
> California Department of Transportation (Caltrans) was called
> in to help clear earthquake rubble from the streets. Media
> personnel reportedly contributed to the crowd problems
> which made it impossible to safely remove the debris. As a
> result, Caltrans threatened to remove its equipment altogether
> unless the traffic could be cleared (Scholl,1984:288; Seismic
> Safety Comm, 1983:70). "The news media was an extremely
> disruptive influence. They frequently hindered response ac-
> tions in their efforts to obtain camera coverage or to interview
> rescue workers, city officials, or other response officials." (Seis-
> mic Safety Comm, 1983:70)

Problems with media air traffic have been reported in a number of disasters
(Tierney, 1985b:34; Seismic Safety Comm, 1983:15; Lewis, 1980:863).

> **EXAMPLE:** *Volcano Eruption, Mt. St. Helens, Washington, May
> 18, 1980.* Private aircraft carrying news media personnel be-
> came a threat to the air search efforts, contributing to the risk of
> mid-air collisions (Drabek, 1981:179; Kilijanek, 1981:63,67).

As a result of repeated problems with air traffic at disasters, the Federal
Aviation Administration was finally forced, in June of 1985, to issue a new
regulation prohibiting helicopters from flying in disaster areas unless they were
carrying out emergency or rescue assignments (Holton, 1985:13).

HOW THE MEDIA OPERATE IN DISASTERS

If disaster planners and managers are going to effectively deal with the media, they first need to understand what makes the media "tick" in disasters. Media behavior is fairly predictable. "Win-win" situations can be accomplished with the news media if disaster planners and emergency managers know what they need and are able to provide it for them (Bernstein, 1986:46).

The Media Will be There

They Will Hear of the Disaster

Emergency managers are rarely able to carry out a disaster response without the media hearing about it (Scanlon, 1985:124). When a major incident occurs, some citizens will call the media. Others will tell friends and neighbors, and that news reaches the media. The media also monitor the activities and radio transmissions of key emergency agencies on a continuous basis. Major response activity is difficult to conceal (Scanlon, 1982:14; Scanlon, 1985:128).

Convergence of Media at the Disaster Site

In a newsworthy disaster, the media may descend on the scene *en masse*. Even a fairly localized disaster can become a world media event within minutes. Literally hundreds of journalists and their crews will show up at the scene (Kilijanek, 1981:77, Wenger, 1985b:7; Scanlon, 1982:14; Scanlon, 1985:124).

> **EXAMPLE:** *Earthquake, Coalinga, California, May 2, 1983.* A swarm of media people arrived—between 150 and 200 . . . " coming "literally from all over the world, taking photographs, interviewing emergency responders and residents, and hiring aircraft to fly over the damaged area. The presence of so many media workers added to the burden of emergency response agencies . . . " (Seismic Safety Comm, 1983:32,34; Tierney, 1985b:34)

> **EXAMPLE:** *Nuclear Reactor Accident, Three Mile Island, Pennsylvania, March 28, 1979.* "The size of the press corps, and especially the electronic legions that descended on the Harrisburg area in the first few days of the crisis were reminiscent in numbers and baggage of the armies of reporters, photogra-

phers, feature writers, correspondents, camera crews and edi-ting and production units that flock every four years to the national political conventions. Only in this case there had been no planning time to establish procedures, or to set up facilities to deal with this overpowering demand for complex informa-tion. Three commercial TV networks . . . established full scale field operations in various motels in the area. . . . Each brought in between 75 and 100 reporters, editors, managers and technicians. In addition, camera and reporter teams were quickly on hand from individual TV stations in nearby Har-risburg, Lancaster, Baltimore and Philadelphia." (Holton, 1985:1)

Effects of Technology on Media Convergence

If a Three Mile Island-type disaster were to occur today, it is estimated that the numbers of media personnel responding in the first 24 hours would be in-creased three-fold. Recent technological advances have reduced the size, weight, and cost of portable video equipment. They have also improved the ability to transmit audiovisual material over long distances by microwave and satellite. Because of this, it has become physically possible and financially feasible for local TV stations to cover news stories in distant locations.

Local TV reporters, complete with their own camera crews, are ranging as far away as Beirut, Jerusalem, Rome, Ethiopia, Peking, India, and Japan to collect lively and graphic material for their broadcasts. This, combined with the intense competition for local news audiences, has led to a multi-fold increase in the number of media personnel who will quickly descend upon a disaster-stricken community (Holton, 1985:3,11). The proliferation of local media at distant events is illustrated by the following:

> **EXAMPLE:** In 1980, the two national political conventions in New York City and Detroit attracted crews from approximately 25 local stations which felt they could afford to send them. In 1984, the New York Times reported that nearly 400 local TV stations were sending news teams to the convention in San Francisco—an 8-fold increase (Holton, 1985:3)!

Information-seeking Behavior

General Types of Information Sought

Members of the press have been trained to report the unique and sensational. In determining the newsworthiness of a disaster, two main criteria are applied

Figure 10-1.   Advances in communications technology have contributed to media convergence at disasters. This media satellite dish was located at "The Fourty-Niner Fire," Nevada County, California, during September, 1988. (Courtesy of *The Union*, Grass Valley, California.)

to an event. The most important is its impact in terms of deaths and injuries; the second is the extent of property damage (Larson, 1980:94,119). Reporters will want to interview participants in the response, officials in charge of disaster operations, witnesses, and victims. Those disseminating information to the media should anticipate requests for the following information (Bernstein, 1986:41; Brunacini, 1978:206.01; Scanlon, 1982:15):

Casualty information.   How many were killed or injured? Of those injured, how serious is their condition? How many escaped? How was escape hindered? Were any of the victims prominent persons? How were the injured managed? Where were they taken? What was the disposition of the dead?

Property damage.   What is the estimated value of property loss? What kind of structures are involved? Did the damage include any particularly

important property (e.g., historical buildings, art treasures, homes of prominent figures)? Is other property threatened? What measures have been undertaken or are being undertaken to protect property? Is the damage covered by insurance? Has this area been damaged by disasters before?

Response and relief activities.
Who discovered the emergency? Who summoned the alarm? How quickly were response units on the scene? What agencies responded? How many are engaged in the response? What acts of heroism occurred? How was the emergency kept from spreading? How are the displaced and homeless being cared for?

Other characteristics of the crisis.
Were there any blasts or explosions? Collapse of structures? Crimes or violence? Attempts at escape or self rescue? What was the extent of the disaster? The duration? Number of spectators? Crowd problems? Were there other unusual happenings? What accompanying accidents have occurred? What were the resulting effects (e.g., anxiety, stress) on the families and survivors?

Causes of the disaster.
Were there any previous indications of danger? Could the disaster have been prevented? How? [Questions about blame are more likely to surface in technological disasters such as hazardous materials accidents] (Wenger, 1985b:22; Holton, 1985:20). Will there be a coroner's inquest? Lawsuits? Insurance company actions? Criminal investigation?

PRINCIPLE
Many of the questions that will be asked by reporters are predictable, and procedures can be established in advance for collecting the desired information.

Types of Media and Their Different Information Needs

In carrying out the public information function, it is important to realize that different types of media have different informational and logistical needs (Scanlon, 1982:17; Scanlon, 1985:127; Holton, 1985:19; Larson, 1980:86; Bernstein, 1986:46; FEMA, 1984a:83,197).

Local versus non-local.
Local news covers all phases of the disaster, ranging from the warning phase, through impact and response, and on into recovery and rehabilitation. The local media have long-range, hometown concerns. They attempt to provide specific information to area residents

to help them face the crisis: warning information, evacuation advice, where help is available, and how long utilities are expected to be out of service.

In contrast, the national media are less interested in details of the disaster, such as names of people (unless they are nationally prominent) or specific locations, than in the unique aspects of a particular disaster or in the human interest stories in the event. They are more concerned with the overall picture, focusing on such things as the scope of the impact, the number of dead and injured, and the activities of federal response agencies and national relief organizations. Their interest in the local incident is generally maintained only as long as there is an active state of disaster. Concern with long-term recovery activity is less likely. Questions posed by the national media may be less sensitive, sometimes trespassing into areas that local reporters consider off limits. (Local reporters must cultivate long-term relationships with local officials who may be important contacts for future stories.) A disaster is more likely to attract national media attention if there is visually exciting footage to film. Events occurring early in the day may get priority for coverage because of the time it takes to edit the video material.

The international press may take on an even different focus. For example, during a flood in Florence, Italy, the British press focused more on the threat to Renaissance art treasures than on the local concerns with the human suffering and loss.

Television.

Television media have concerns with shipping film or transmitting video material via satellite. TV combines the qualities of sight, sound, immediacy, and motion. The impact of television is affected by the editing of the stories, the hour of broadcast, and the number of times the broadcast is repeated. The TV media want visuals; a telephone conversation is not an adequate substitute. Television is predisposed to a headline approach that emphasizes succinct, catchy 20- to 30-second statements. The haste required in assembling a major television news report is illustrated in a passage by ABC News vice president, A.V. Westin, in his book, *News-Watch*, as described in the Federal Emergency Management Agency's Monograph on the electronic news media:

> "In its live and continuous coverage of a breaking story of . . . significant proportions TV news compresses that [traditional editing] process. The coverage takes on a life of its own, developing momentum and drive, which forces correspondents to race rather than walk from one story element to another. The anchors become, in effect, editors and reporters. Reporters on the scene of the breaking story may add important details; still it is the anchor who summarizes, repeats, amplifies, and ultimately evaluates the material coming in from the field. All

those judgments are taking place under time pressures and in the midst of near chaos. Dozens of facts, rumors, conjectures, and ideas are surging about every minute. The system does not provide for a detailed review of material. Much of what pours in are raw data, edited in the head of the correspondent as he or she reads from notes hastily scribbled at the scene. Interviews are done with eyewitnesses whose own credibility is unchallenged and unchecked. The system provides for only two alternatives: accept the material or reject it." (Holton, 1985:13)

Actually, to fully appreciate the impact of dealing with a major disaster story, what is described in the above passage would have to be mentally multiplied fifty-fold (Holton, 1985:12).

Commercial radio.   Radio wants to be there first and to have rapid access to information. After all, it is able to reach audiences almost anywhere. It tends to broadcast the information almost as soon as it's received, but generally limits its reports, which are short in duration (often less than a minute) and selective in detail.

Print (newspapers, magazines).   Radio and print media are concerned with availability of telephone communications to transmit information to their offices. Because print news does not have the time constraints experienced by radio and TV, it tends to search for more background and off-beat material. Often reports include analysis and commentary, and stories may build day-after-day as the disaster progresses. They want depth and graphics and are not constrained by the same time limitations as radio and TV. Print photographers may be very aggressive, because to compete with the emotional impact of TV, their pictures have to be exclusive.

Information-sharing

One of the important maxims of the media is that while it is desirable to get an exclusive story (a "scoop"), it is even more important not to *get* scooped. It has been observed that the tendency in a truly major incident is for almost all the reporters present from the various media to share with each other the information they obtain. The various media also monitor each other to pick up information they may have missed (Scanlon, 1982:15; Scanlon, 1985:131; Wenger, 1985b:20; Holton, 1985:21).

EXAMPLE: The Canadian public broadcasting organization (CBC) receives information from Canadian Press (CP), the main Canadian news agency. CP is linked to the United States,

British, and French news wire services, with which it shares information. CBC is also linked to U.S. television networks ABC and NBC. In addition, CBC monitors its main rival, CTV, and the Toronto newspaper, the *Globe and Mail* (Scanlon, 1982:15).

This information-sharing tendency has important implications for the establishment of a centralized source of public information in disasters (discussed later in this chapter).

PRINCIPLE

Newsworthy information will rapidly spread among news organizations and from one type of media to another.

The Media are Different in Disasters

Diminished Information Verification

In disasters, the media become voracious consumers of information. Television and radio stations may abandon regular programming in favor of non-stop disaster coverage. News shortages develop when official sources of information are not sufficient to fill the expanded news coverage. Sometimes the pressure to keep airtime filled with disaster news creates demands on media personnel at the scene that can only be described as desperation. When this occurs, the usual pattern of checking sources is often relinquished in favor of a new norm: that all news should be disseminated from all sources (official and non-official) as soon as it is obtained. Even contradictory information may be broadcast. This is justified on the assumption that instant feedback from the public will straighten things out in the end (Larson, 1980:62; Holton, 1985:20; Scanlon, 1982:15; Scanlon, 1985:128).

Diminished Adversarial Role

The traditional role of the press as a watchdog over the government goes back to the very founding of our nation. However, in disasters, the media will often moderate their adversarial posture towards government. Quite often, the media will arrive at the scene willing to temporarily set aside animosities that might have developed during routine news coverage (Holton, 1985:22). Although this norm has been applied to technical disasters as well as natural ones, this is less likely. In particular, nuclear accidents tend to arouse a skeptical stance and blame seeking. Whether justified or not, some journalists feel the nuclear proponents have created a credibility gap when it comes to the safety of power plant reactors. In addition, a significant anti-nuclear movement generates a hostile and untrusting stance towards emergencies involving radioactive material (Rubin, 1987:14).

The media will often cooperate with requests by officials to hold back information that might have an adverse effect on the public during the disaster. Even if there is no such request, information has been withheld on the media's own initiative when it was felt to be harmful.

> **EXAMPLE:** *Nuclear Reactor Accident, Three Mile Island, Pennsylvania, March 28, 1979.* The Harrisburg *Patriot-News*, acting solely on its own judgment, deleted a reporter's account of what would happen to bank records in the event of a disaster because it was concerned about provoking a run on the banks (Scanlon, 1985:124).

PRINCIPLE

The media will often withhold newsworthy disaster stories it feels would be detrimental to the public.

Domination by the National versus Local Media

In routine emergencies, response and government officials may deal with familiar persons representing the local media. To some extent, local reporters have a stake in cultivating and maintaining good contacts with local officials who are often a reliable source of daily news. In addition, local reporters have some sensitivity about the needs of the community. These factors may allow a good working relationship to develop between local media and response organizations.

Some officials may come to believe that this relationship will form the basis of their interactions with the media in a major disaster. Unfortunately, the usual ground rules of interaction are often brushed aside by the outside media whose numbers quickly overwhelm the locals. In particular, network television news tends to quickly dominate at the scene of a major disaster. There are several reasons for this (Holton, 1985:21):

- The networks assign large, multi-specialty staffs to disasters, often led by producers and other personnel experienced in this kind of event. These crews are often quite adept at acquiring lodging, office space, telephone lines, and such.

- The major networks are pragmatically aware of the advantages of cooperating with each other and have developed arrangements for doing so effectively. The effectiveness of such agreements is the basis of an efficient news-gathering operation.

- Usually each network assigns several "star" correspondents to an event of such proportions. The possibility of an appearance on "Good Morning America" or "Dan Rather's Evening News" tends to sweep aside the local media competition for interviews.

PRINCIPLE
Local officials will have to deal with different news media in times of disaster than those with which they interface on a routine basis.

The "Command Post" Perspective of the Media

For all types of media, the most important sources of news are official government agencies, and much of the news about disasters tends to be reported from the perspective of these agencies. This is known as the "command post" perspective (Larson, 1980:89; Quarantelli, 1981b). In fact, the media can be depended on to demand news conferences at which authoritative official statements can be recorded (Scanlon, 1982:16). This, like the tendency for reporters to share information, can facilitate response agency efforts to develop a centralized source for public information.

Media Difficulties with Technical Information

Most news agencies have not developed a cadre of reporters and consultants with technical expertise on disaster-related topics. Accordingly, they have a great deal of difficulty evaluating the technical aspects of disasters and assessing the competence of various (sometimes conflicting) sources of information (Drabek, 1986:167; Scanlon, 1982:16; Scanlon, 1985:124; Wenger, 1985b:22; Holton, 1985:24).

> **EXAMPLE:** *Nuclear Reactor Accident, Three Mile Island, Pennsylvania, March 28, 1979.* "A seemingly simple question of whether the core of the reactor had been . . . damaged elicited

responses couched in terms of 'ruptured fuel pins,' 'pinholes in the cladding,' 'fuel damage.' " (Scanlon,1985:124) Once a reporter had figured out what they were saying he might discover another source that was saying something different. Media representatives felt as if they needed a degree in nuclear physics just to ask the right questions (Scanlon, 1982:17).

This factor makes it very important for public information officers to find those to speak to the media who are adept at translating technical subjects into plain English.

IMPROVING MEDIA DISASTER OPERATIONS

The Need for Media Planning

Considering the impact the media can have on the public and disaster response agencies, both positive and negative, the need for planning in this area is of paramount importance.

> **EXAMPLE:** *Train Derailment and Chlorine Gas Leak, Mississauga, Ontario, Canada, November 10, 1979.* The evacuation of 220,000 residents, 3 hospitals, and several nursing homes was widely praised. The success of the response was attributed, at least in part, to the fact that the area police put the importance of media operations second only to on-site command and control (Scanlon, 1985:124,127; Quarantelli, 1982a:H-36).

However, those who have studied this topic have reported few communities that have treated this facet of preparedness as more than an afterthought. Often this has consisted merely of assigning someone the duty of writing and distributing press releases and assuming this would adequately deal with the issue. It has been claimed that a number of the problems disaster managers face in dealing with the media result from a failure to fully understand and plan for the media. Even those who understand what information the media needs and the importance of providing it are not always provided the resources (e.g., staff, time, equipment) to accomplish the task (Wenger, 1985b:8,25; Holton, 1985:iii).

Ability of the Media to Survive and Function

If the local media are to be able to carry out their designated functions in a disaster, they must first be able to survive the impact. Consideration must be given to the importance of locating broadcast stations and transmitters in areas not vulnerable to natural hazards and to providing back-up electrical generators.

> **EXAMPLE:** *Flood, Rochester, Minnesota, 1978.* When flood waters crested the Zumbro River and inundated the city, television station KTTC was in the water's path. Local news coverage ended when the station crew was forced to evacuate. Because the transmitter was located upstairs, the crew was able to leave the microwave link operating so that at least network broadcasting could continue. However, even this was terminated when the local power station flooded, cutting off the station's electricity. Not only was the station unable to function as it should during the disaster, but the waters also destroyed most of the film and video tapes in the basement on which were stored 20 years of Rochester history (NAB:1).

It is also important that communication links to sources of official information not depend on telephone lines. Additional considerations include provisions for sleeping and eating, for calling in staff after hours, and for rotating staff on a 24-hour basis (Scanlon, 1985:124).

Educating the Media about Disasters

As in the case of disaster response organizations, media function is best carried out when the participants have an adequate disaster knowledge base. Education can help to reduce inaccurate news reporting. For example, newscasters should be encouraged not to withhold news information, warnings, and instructions to the public for fear of causing panic. The evidence indicates that the public is quite capable of handling the facts. The media should be aware that a disaster warning is less likely to be taken seriously if it is followed by resumption of normal programming. Newscasters should also be taught the importance of announcing the areas *not hit* by the disaster and the effect this information will have on reducing the number of calls by persons who believe they have loved ones in the impact zone. Finally, reporters should be sensitized to the fact that precise information and accurate figures on deaths, injuries, damage, and cause, in the early aftermath of a disaster, is an unrealistic expectation. Trying to pin down officials to "exact" information can be self-defeating (Scanlon, 1985:129).

Media organizations often keep files to which they refer for background information when a large news story breaks (Scanlon, 1982:14). This provides an opportunity for astute disaster planners to make useful and accurate information available on a pre-impact basis. Glossaries of technical terms, schematic drawings, models, and diagrams are also valuable tools for assuring that the media, and ultimately the public, understand disaster phenomena (Bernstein, 1986:86).

Involvement of the Media in Planning

Probably one of the best ways to educate the media about disasters is to have them involved in the disaster planning process. Effective planning for public information in disasters cannot be done *for* them. It has to be done *with* them (Wenger, 1985b:25). Without their active involvement, one suffers the risk of falling into the "paper" plan syndrome.

Representation on Planning Bodies

One of the most important requirements for effective media disaster operations is to plug the media into all stages of the disaster planning process (Scanlon, 1985:126). The media should be represented on disaster-planning bodies. Some may oppose this idea believing that it may lead to a hesitancy to discuss preparedness weak points in the presence of the press. In that case, it may be necessary to be selective about the meetings the media should attend. On the other hand, the media can be a powerful ally when it comes to obtaining public funding and support if they understand the weak points and see a need for financial bolstering.

Local Media Participation in Media Relations

One clever strategy for handling media relations is to delegate the responsibility to the local media itself. This technique has reportedly been used effectively in a number of Canadian communities. The media has been plugged right into the disaster response and served as a liaison between the Emergency Operations Center and the outside media (Scanlon, 1985:127).

Initiating Planning with the Media

When initiating contact with the media, it is probably best to start at the top. Contact the commercial radio or television station owner or manager, the newspaper or news magazine publisher or editor. They will be able to assign someone in their organization to act as liaison with the local disaster planning effort. An alternative is to directly approach those responsible for day-to-day news coverage. For the print media this would be the managing or city editor, for the electronic media, the news director (Scanlon, 1985:126).

PRINCIPLE

Adequate disaster preparedness requires planning *with* the media rather than *for* the media.

Reaching the Public

Knowledge about Audience Composition

Merely seeing to it that there is a local radio or TV station broadcasting disaster information does not assure it will be received by all who need it. For example, some communities have significant populations of special language or ethnic groups that may not be reached by English-language broadcasting. Those with hearing or visual disabilities have to be considered as well.

A helpful technique in public information planning is the use of viewing and listening audience surveys carried out by agencies such as A.C. Nielson and Arbitron. These surveys can be used to provide information about the audiences for various media at different times. By inference, this information can also be used to get an idea of who is not being reached (Scanlon, 1985:125). One consideration may be to use volunteers, or "TelePatrol" (see Chapter 9) to canvass neighborhoods to ascertain who cannot be reached by the normal warning and advisory methods, and to develop alternative means of conveying this information when needed.

The Emergency Broadcast System

The Emergency Broadcast System (EBS) evolved from the CONELRAD (Control of Electromagnetic Radiation) system created in 1951. Its purpose was to provide a means by which the President, utilizing existing commercial broadcasting stations, would communicate directly with the public in times of national emergency. While the system was originally to be used in the event of enemy attack, it has been expanded to civil disaster use so that it can be activated by certain state and local government officials as well (NAB:20).

State-wide activation of the Emergency Broadcast System allows the governor to address the entire state citizenry at the same time. However, the system has an inherent technical weakness. It depends on "chain-broadcasting." What this means is that the governor broadcasts a message via a commercial (usually an FM) station in the Capital. This broadcast is picked up by a more distant station and relayed on. This process is repeated until the message is conveyed across the state. If, however, one station in the chain is disabled, or can't operate because it doesn't have an emergency generator, then the "chain" is broken.

Attempts to use the EBS can fail, as in after the earthquake in Coalinga, California, when officials didn't know how to activate it (Seismic Safety Comm, 1983:165). Unless government officials and commercial broadcasters are motivated enough to educate the users and test the system frequently, those who need to utilize the system in a disaster may be unfamiliar with it (another variation of the "paper" plan syndrome) (Harter, 1985).

Exceptions to FCC Regulations

In emergency conditions and disasters, the Federal Comunications Commission (FCC) allows public and commercial broadcast stations, without prior approval, to ignore certain regulations governing their transmissions. For example, certain stations have to go off the air or limit their transmission power at certain times of the day or night. During an emergency, stations may use their full power regardless of the time of day. Furthermore, at the request of government officials, stations may broadcast point-to-point and even person-to-person messages for the purpose of requesting or dispatching aid and assisting in rescue operations (NAB:13; Wenger, 1985b:18).

Use of Non-local Stations

When local stations are non-functional as a result of a disaster, it is often possible for local media or local government officials to convey information to stations outside the impacted area for broadcast back into the stricken locale. In this way, the flow of essential information to the disaster victims can be maintained.

TIS Radio: Be Your Own Media

Using a little-known provision under Part 90 of the FCC regulations, agencies of city, county, and state government are able to set up their own "mini" radio stations. These "Travelers Information System" (TIS) radio stations, can transmit information to the public on the AM radio band. You may have encountered such a radio station while visiting a national park, a major airport, or while crossing an international border. Typically, you will see a sign at the roadside instructing you to tune to either 530 or 1610 Kilohertz on the AM band for local traveler's information.

TIS radio stations can be licensed for use in national parks, adjacent to major transportation centers, and adjacent to federal and state highways. It is this latter provision (being adjacent to a highway) that allows almost any city to qualify for a TIS radio station license. While most TIS stations are used to provide traffic directions and information, they can also be used for other non-commercial purposes, including the issuance of public safety information during emergencies and disasters.

The Coronado, California, Police Department established a TIS radio station in January, 1985, and has found it to be a worthwhile and cost-effective addition to their public information system in routine and emergency conditions (Boyd, 1985). Motorists entering the city, will see large road signs reading, "Traffic/ Emergency Info—Tune 530 AM," and can get up-to-the-minute information on traffic problems and delays. They may be directed to alternative routes to avoid traffic congestion, relieving traffic officers for other purposes. During local emergencies or disasters, the radio can broadcast official information on the event, lessening the volume of telephone inquiries that typically "light up" police telephone switchboards in such events, and allowing personnel to direct their attention to emergency response-related communications.

The equipment for a TIS radio station is relatively inexpensive ($1,500 to $2,500), small in size (the transmitter occupies a space of about a cubic foot), and simple to operate (estimated training time is less than 5 minutes). Endless-loop audio tape cassettes are available in various lengths (20 seconds to 10 minutes and more) for about $5 each.

Total installation time for the system is about 4 hours. Technical assistance can be obtained from local government radio maintenance personnel, or donated time from local AM commercial radio stations. In Coronado, assistance was given by local amateur radio operators who were already involved in local disaster preparedness activities. Coverage area is dependent on the type of antenna, its placement, the local terrain, and the maximum allowable transmitter power, which is limited to 10 watts. Even with these limitations, however, the broadcasts can be received within a radius of 7 to 8 miles.

The steps needed to establish a TIS radio station are as follows:

- Obtain a TIS radio station license application and a copy of Part 90 of the FCC rules. Request information regarding commercial radio stations in your area that broadcast between 520 to 540 kHz and 1630 to 1640 kHz. You must operate on a non-interference basis with other local stations.

- Using available or volunteer technical assistance, prepare the FCC license application. You can expect the license to arrive in 60 to 90 days.

- Locate a radio equipment supplier. Several manufacturers produce TIS radios. Advertisements can be found in communications journals (e.g., *The Associated Public Communications Officers* [APCO] *Journal*, P.O. Box 669, New Smyrna Beach, FL 32069). Delivery time will be 4 to 8 weeks. Two manufacturers are:

Radio Systems, Inc.
5113 W. Chester Pike
Edgemont, PA 19028
Attn. Dan Braverman
(800) 523-2133
or (215) 356-4700

Information Station
Specialists
P.O. Box 51
Zeeland, MI 49464
Attn. William Baker
(616) 772-2300

- Using available technical expertise, identify a location for the equipment. You will need a 115 volt AC circuit which is on emergency backup power. A suitable location for the antenna must be located, consistent with FCC regulations which limit the height of the antenna above the average local terrain. The bulk of the installation time will be installing the antenna, wiring it to the transmitter, and tuning the antenna to match the radio.

Centralization of Public Information

Having a central source from which the media can receive official information about the disaster can help to assure that what the public receives is timely, consistent, and accurate. The Incident Command System has seen this as an important enough issue to designate a specific command staff position for it and to develop specific procedures for the process (ICS, 1981). Indeed, the design of the ICS and the forms used to keep track of the incident and resource status, contribute to the effective and efficient collection of information for the media.

Acceptance by the Media

Several characteristics of the media make them receptive to a central source of public information:

- The media tend to share information anyway.

- The media often prefer to attribute the news to official sources (the "command post" perspective).

- Due to news deadlines, the media will congregate where it is easiest to get the greatest amount of news quickly.

- The media will not only be receptive to press conferences in a central location where "packaged" news releases are handed out—they will often, in fact, demand them (Scanlon, 1985:124).

Where the concept has been promoted, the media, especially television, have given it support and even offered to help preplan the details of such an arrangement (Holton, 1985:27).

The need for a public information center, as seen from the perspective of the media, is captured in the following quote:

> **EXAMPLE:** *Tornado, Worcester, Massachusetts, June 9, 1953.*
> "Whom do you believe when you are broadcasting? Who is the real authority? There is a confused picture from confused lines of authority. They are superseding, overlapping, and parallel to each other, all at the same time. . . . But who knows the facts and can speak with authority about them? That is the dilemma of the broadcaster. . . . That is why a central clearing house of information is important. . . . When there is just confused authority, the broadcaster has to do the best he can. He has to make private decisions about what he will regard as authoritative sources." (Rosow, 1977:86)

Sources of Information

The success of a centralized information office depends on its access to accurate and timely intelligence (situation analysis) (Rubin,1987:12). This requires functioning procedures for inter-organizational information flow, a problem discussed in detail in Chapter 5. The adoption of the Incident Command System is one way of facilitating this process.

PRINCIPLE

The propensity for the media to share information and to assume the "command post" perspective facilitates the establishment of a central source of disaster information.

Limitations of Centralized Information

The fact that the media will embrace the establishment of a central source of disaster information does not mean that any emergency management or government official is going to be able to control the news. Reporters do not *rely* on only one source for news information. They usually have a variety of contacts and sources, and this is not likely to change in a disaster.

Although it is advantageous to establish a single media center for disaster information and to staff it with trained public information officers, the amount of information that is centralized is relative, rather than complete. To some extent, rumors will still occur, the public will still get information from friends, family members, and other sources, and the credibility of government officials will not be guaranteed (Wenger, 1985b:21).

Reducing Media Convergence

Media Pools

The problem of media convergence—crowding the emergency operations center, tying up communications facilities, and making demands for tours of the disaster site—can be lessened by the use of media pools. A media pool is where only one or a few reporters collect information about an event and then share it with their colleagues and competitors. The media are often not only receptive to pools, but (especially network television) will even form pools on their own initiative (Holton, 1985:21).

One tactic that might be useful is to assign local media representatives as "pool coordinators." The selection of these coordinators would depend on their taking a disaster public information training course set up by the local preparedness agency. This "disaster education" can help avoid the dissemination of disaster myths, can help the media to understand and sympathize with the difficulties faced by public officials, and can help the media to be aware of the types of public information that best assist the disaster response efforts.

Because of the differences in media needs, it is best to compose the media pool of representatives from each of the following: print media, television, commercial radio, and the national wire services. In addition, the pool should include representatives from local, regional, national, and if applicable, international media (Bernstein, 1986:88).

Media Outreach and/or Use of Computer Communication

Another way to reduce the media's disruptive activities is to channel their information-seeking behavior. This can be done by contacting the various media and informing them where centralized information can be reached. Don't wait for the media to show up—get the information to them where they are, and use automated or computerized devices to give regular news updates.

The mere existence or designation of a centralized information office does not guarantee that it will be used. The media must be aware that it has been set up and where it is located. If media pools have been established, then media outside of the pool need to know how to obtain information from it.

Printed press releases or pre-recorded (audio and/or visual) press releases can lessen the burden on public information officials. Releases can even be made available by recorded telephone messages. In fact, TelePatrol (see Chapter 9) could be well adapted for this use. Specifically tailored recorded messages can be assembled for the different types of media and telephoned to them by a computer dialing process using a pre-programmed list of phone numbers. Centralized sources for visual material can also be conveyed in the messages.

Channeling the news media's information search is best accomplished, if possible, by contacting them before they show up at the scene. Once again, if telephone lines are available, this may be accomplished by TelePatrol. Another arrangement is to plan a tone-encoded, two-way radio net with local news media and local or regional wire service offices.

One program that illustrates the usefulness of the "media outreach" idea is the *Media Computer Network* (MCN), in use in Sacramento, California (Dickson, 1987). The county sheriff was the motivating force behind the idea. The network uses computer terminals in the public information offices of a number of public safety and other agencies which are tied by dedicated lines directly into area news agencies. At present, the network includes local TV and radio stations, newspapers, the local and Associated Press wire services, area police and fire agencies, the County District Attorney's office, the California Highway Patrol, California Office of Emergency Services, Folsom Prison, the local offices of the FBI and U.S. Attorney, and local utilities including the Rancho Seco nuclear power plant.

Each terminal can send, receive, and store messages. High-quality graphics printers allow maps, photos, and other visual material to be transmitted. Portable, briefcase-sized, inexpensive computers are available for agencies to use in the field, and cellular phones and phone booths can be used to send information from them. When a local agency, such as the police department, types a news bulletin into the system, it is immediately and automatically conveyed to every other terminal in the system. In addition, individual terminals can communicate with each other privately.

Experience with the network has revealed several benefits. First, local agencies find that use of the system decreases the number of inquiries by the press. Since the network conveys information instantly and simultaneously to 22 separate news agencies, including two wire services, this lessens the need for each of these agencies to make a separate inquiry. Prior to initiation of the network it was not unusual for the law enforcement agency to receive 20 to 25 calls on a routine developing crime story. In other words, there reportedly is a noticeable decrease in media convergence. Second, the system is in use on a daily, routine basis. For this reason, in contrast to the Emergency Broadcast System, it is more likely to be a familiar tool and more likely to be functioning and to be used in a disaster. Third, since every agency receives a printed copy of the news releases, information is less likely to be misunderstood, and incorrect information is more likely to be noticed and corrected by somebody on the network.

Although MCN is a proprietary program, and the author does not wish to endorse specific commercial products, the idea appears to have merit. Readers wishing to investigate the idea further may contact: The Media Computer Network, P.O. Box 60919, Sacramento, CA 95860, (916) 488-8624.

Regional Public Information Teams and Equipment

Another novel approach that has been suggested for managing the media assault in disasters is to import specialized outside public information management equipment and/or teams of public relations specialists. Media coverage in disasters is dominated by the national television media. Perhaps it would be good to make available to local governments teams of professionals who are experienced in handling the media (Scanlon, 1985:131).

Surviving a Press Conference or Interview

To the harried official trying to manage the pressing contingencies of a major disaster, the idea of facing a national news audience can be an intimidating prospect. In fact, the absence of skill in carrying out this task can convert an otherwise adequately handled incident into a public relations disaster. What follows, while certainly not scientifically validated, is a collation of suggestions from public information professionals which may be of help in surviving the experience (FEMA, 1984a:88,198; Bernstein, 1986:45,86; Lavalla, 1983:279; Scanlon, 1985:127; Johnson, 1986:106; CSTI, 1987).

Be Prepared

There are certain questions (discussed above) that you can virtually predict the press will ask. Do your homework and be ready with answers. Background material, graphs, charts, and illustrations can sometimes be assembled in advance to help convey what you will want to say.

Be Wary of "Off-the-record" Comments

The reporter assumes that everything you say and everything he sees is "on-the-record" and fair game unless specifically agreed to the contrary in advance. It is wise, therefore, when talking to the press, either in person or on the phone, to assume that anything you say might be published. If you feel the information you provide should be anonymous, you might be better off not giving the interview at all.

Be Honest

Trying to cover up mistakes, mislead the press, or withhold critical information about a disaster can backfire.

> **EXAMPLE:** *Nuclear Reactor Accidents, Three Mile Island, Pennsylvania, March 28, 1979 and Chernobyl, USSR, April 28, 1986.* "Once officials in charge did begin to talk, they were quick to put the best face on developments and reluctant to confirm bad news. This diminished their credibility and severely reduced the number of trustworthy sources with firsthand knowledge of the accident. Similarly, they provided certain kinds of information—particularly about radiation release—too late to help a worried public. This compounded their credibility problem." (Rubin, 1987)

When reporters find officials reticent to give accurate and timely information, they may impute the worst motives for the apparent evasiveness and seek alternative (possibly less reliable and authoritative) sources to get the "facts." Sometimes this also prompts them to consider the worst-case scenario in describing the disaster and the response effectiveness. This is particularly the case in controversial or politically charged issues such as nuclear accidents. Reporters have had to deal with public relations and government spokesmen who are experts at verbal misdirection. The misrepresentation and deceit attributed to government statements about events like Watergate and Vietnam have led reporters to develop a healthy skepticism when they encounter evasiveness and "no comment" responses. As a result, such statements are generally interpreted to indicate that the interviewee either doesn't know the answer or that he has something to hide (Rubin, 1987; Johnson, 1986:111).

There are, of course, times when it is legitimate to hold back specific types of information. When this is the case, it is best to state in a matter-of-fact way why the information is being withheld and when it might be made available, if that is possible. If the media have obtained information that could be dangerous to release, do not hesitate to ask them to withhold it. Be sure to give them the reasons why it should not be reported and promise to tell them when it is safe to release the information.

Admit problems and mistakes if they exist. Significant errors that are concealed will leak out and cause much more difficulty than if they had been immediately disclosed. On the other hand, this does not mean that officials are obligated to cast everything in its worst possible light. If problems have developed or mistakes have been made, point out your positive efforts to correct them and then turn the conversation toward what has been done to effectively manage the emergency.

Stick to the facts. Don't describe the situation as any better or worse than it actually is. Avoid making statements that could be construed as trying to exaggerate or to grab undue credit or manipulate the story to create a favorable impression of your office or agency.

Manage Ambiguity

Disasters are characterized by uncertainty. Often, accurate information—about the extent of the destruction, dangers to public safety, cause of the incident, and other matters of interest to the media—is simply not available early on. The interviewee should be able to admit what he doesn't know, and avoid speculating on the answers. Instead, he should state that he will try to determine the facts as soon as practical and make this information available.

Relate to the Audience

Remember that you are not talking to an audience of fellow experts. Avoid using technical terminology and jargon. This only confuses the public or makes them think you are trying to hide something. On the other hand, avoid "talking down" to the audience too. Be human; strive for an informal, conversational tone, while still maintaining a professional demeanor.

Take the Initiative

Often, the interviewee is chosen because he is the expert. As such, he, and not the reporter, is in the best position to judge what the important issues are. The interviewee should take the lead in pointing these out and directing the course of discussion. This can be accomplished even in the worst possible press confrontation, the "ambush interview." When the reporter suddenly pushes the microphone into your face and asks a difficult question, you can use what is called "transition technique" to redirect the direction of the interview. This is done by briefly answering the question (e.g., "yes, this is true, but . . . " or, "No, that is not the case, and . . . ") then proceeding to address what you consider the important facts. The interviewee should say what he needs to say, not just answer the questions posed by the reporter.

The interviewee should ask the reporter to rephrase the question if what is being asked is not clear. ("Did you mean . . . ?") Do not allow misleading or inaccurate comments or statements to go unchallenged. Correct any bad information built into the questions. The longer misinformation lingers before being corrected, the more it gains credibility. Refute it politely, offering a brief explanation, and move on to the topics you want to cover.

The Team Approach

Sometimes fear of the media can be the interviewee's worst enemy. One way of tackling this problem is to avoid facing the interview alone. Instead, assemble a group of experts and spokespersons and face the media as a team.

Preparation for a Television Appearance

While preparation for any interface with the media is important, television requires special attention. Public relations professionals describe television as an emotional medium, rather than an intellectual one. They say that, while viewers will often forget the content of your message, they will remember your style: how you looked, your manner, and the quality of your voice. First impressions are important, and there is an advantage to having your most important and most positive statements up front. Conventional clothing should be worn (as well as accessories, jewelry, makeup, and hairstyle) with subdued colors and with a design that reflects your professional image. If it is possible to choose in advance, clothing colors that blend into the set background should be avoided.

As noted before, television relies on a "headline approach" with succinct, catchy, 20- to 30-second statements. The interviewee can almost certainly assure which part of the pretaped material will be used by prefacing it with, "The most important thing about this is. . . . " Most newscasters work under tight deadlines and appreciate when the experts flag what is important for them. While maintaining accuracy, try to keep answers short and to the point. The best way to lose what you are trying to get across is to "overload the system" by giving too much information.

Telephone Interviews

Find out to whom you are speaking, so your answers can best meet the needs of the interviewer. Also get a phone number so you can call back with additional important information or corrections. Find out when and how the material is going to be used and who the target audience is. Be sure you have a clear telephone connection. Unfortunate misquotes can result because a statement was not clearly heard. Have your interview statements read back to you. Start off by stating your main point in clear, simple language and repeat it in a concluding summary statement. Subtle or wry statements may be translated badly—avoid them. Also avoid making absolute statements. If you are in doubt about an important point, or wish to give it further thought before answering, tell the interviewer you will call back shortly—and be sure to do so. Also be sure to call back if new information develops which causes you to want to change the statements you made or the opinions you gave. Offer any photographs, charts, or illustrations that might contribute to the story if the report is to appear in the print media. Try to have some good background reference material available for reporters who are not familiar with your subject.

If the telephone interview is being taped for radio, you may have to do a few trial runs before getting an adequate product for broadcast. Be sure to find out what the time limits of the broadcasted interview are so you can avoid giving too much information. It is best that you, as the expert, determine what is most important, rather than having the reporter edit a few seconds of material out of a 20-minute interview. Turn off any noisy air conditioners or other equipment in your office that might interfere with the sound quality of your statements. Have other phone calls held. Ask whether you should use the interviewer's name in your responses. If you pause to think, avoid "ah's" and "uh's;" it is better to have a silent pause. Avoid the use of numbers; if they are essential, however, round them off and use as few as possible. Talk in a normal tone *past* the telephone mouthpiece rather than directly into it. Don't hold the mouthpiece too close. This will help prevent popping and hissing sounds when pronouncing "p," "s," and "t" sounds.

The Interviewee's "Bill of Rights"

Although providing the media with accurate, consistent, and timely information can often help an organization carry out its function, maintain a good public image, and help the public, the potential interviewee does not always have to place himself completely at the beck and call and at the mercy of the media.

Organizations do have the right to remain silent, to stick to a prepared text, or to say, "I don't know. I'll find out and get back to you later." They also have the right to tape the interview, to keep records of the information given, to have others present, and to use their best spokesperson to face the press. If faced with a notorious, cut-throat interviewer, one tactic that might be useful is to consent to an interview, but only if it will be reported verbatim and in full (except for editing that is reviewed by and mutually agreeable to interviewer and interviewee).

SUMMARY

The need for a free press is an important part of our heritage. Nonetheless, disaster managers often feel that they could do their job better if they didn't have to divert valuable time, resources, and effort to deal with the press. In many communities, the establishment of procedures for effective media relations in disasters is not given high priority. Preparation may consist merely of assigning someone the responsibility for handing out press releases and talking to reporters that show up.

Media behavior in disasters is, to a significant extent, predictable. Because of this, proper public information procedures are likely to reduce the disruption caused by media demands. In this chapter we have examined the types of behavior that can be expected of the media in disasters and discussed several techniques for dealing with them.

PLANNING CHECKPOINTS

[ ] Is the mass media represented on your area's disaster planning body?

[ ] Do your local media organizations have emergency backup electrical power and other provisions allowing them to function in a disaster?

[ ] Are the facilities of your local media located in areas vulnerable to disaster threats (e.g., in the flood plain)?

[ ] Do your local TV and radio stations have pre-established procedures and agreements for conveying information to non-local stations for broadcast back into the area in the event that local stations are knocked off the air by a disaster?

[ ] Do your local broadcast stations know what FCC regulations do not apply in disasters?

[ ] Do you have a program to educate the local media about disasters?

[ ] Do you have provisions to provide background and public educational and instructional information to the media for use in the event of a disaster?

[ ] Does your area test the Emergency Broadcast System, including the interface between local government and local EBS stations, on a monthly basis? Do local officials know how to activate and use the system?

[ ] Do your local emergency response organizations have a mutually agreeable procedure for centralized disaster information dissemination? For collecting information from the various response organizations for the information center (e.g., the procedures outlined in the Incident Command System)?

[ ] Does your public information plan include the use of media pools?

[ ] Do your emergency managers have procedures and guidelines to follow in anticipation of information the media will request? Have provisions been made to provide it to them in a way that is minimally disruptive to emergency operations? Do they *initiate* contact with the media to provide this information? Are automated devices used to provide information to all local media (e.g., a computer electronic mail system)?

[ ] Has your area considered the establishment of a TIS radio station?

ADDITIONAL READING

Bernstein AB: The emergency public relations manual, revised edition, 1986. Available from: PASE, Inc, PO Box 1299, Highland Park, NJ, 08904, $125.00.

Committee on Disasters and the Mass Media, Commission on Sociotechnical Systems, National Research Council: Disasters and the mass media, 1980. Available from: National Academy of Sciences, 2101 Constitution Ave, Washington, DC 20418, $9.75.

Friedman B, Lockwood D, Snowden L, et al: Mass media and disasters: an annotated bibliography, Miscellaneous Report 36, 1986. Available from: Disaster Research Center, University of Delaware, Newark, Del, 19716.

Hartsough DM and Mileti DS: The Media in Disaster, In: Laube J and Murphy SA, Perspective on disaster recovery, p. 282, Appleton-Century-Crofts, Norwalk, Conn, 1985.

Holton JL: The electronic media and disasters in the high-tech age, publication 109, 1985. Available from: Federal Emergency Management Agency, PO Box 70274, Washington, DC, 20024, free.

Johnson B: Dealing with Mass Media in Time of Emergency. In: Multiple death disaster response workshop, p 104, Publication SM/IG 193, 1986. Available from: Federal Emergency Management Agency, PO Box 70274, Washington, DC, 20024, free.

Lavalla R and Stoffel S: Managing External Influences. In: Blueprint for community emergency management: a text for managing emergency operations, p 273, 1983, Available from: Emergency Response Institute, 1819 Mark St, NE, Olympia, Wash 98506, $25.00.

Policy Development and the Media. In: Formulating public policy in emergency management, Course Book and Resource Manual, Student Manual 51, p 82, 1984. Available from: Federal Emergency Management Agency, PO Box 70274, Washington, DC, 20024, free.

Scanlon J and Alldred S: Media coverage of disasters: the same old story, Emergency Planning Digest (Canada), 7(4):13, Oct.–Dec. 1982. Available from: Emergency Preparedness—Canada, 141 Laurier Avenue West, 2nd floor, Ottowa, Canada, K1A OW6.

Scanlon J, Alldred S, Farrell A, et al: Coping with the media in disasters: some predictable problems, Public Administration Review, 45(Special Issue on: Emergency Management: A Challenge for Public Administration):123, (Jan.) 1985, Available from: the American Society for Public Administration, 1120 G St, NW, Washington, DC 20005.

Wenger D: Mass Media and Disasters, Preliminary Paper 98, 1985. Available from: Disaster Research Center, University of Delaware, Newark, Del 19716.

APPENDIX

A

HAZARD ASSESSMENT

THE NEED FOR HAZARD ASSESSMENT

Disaster planning is carried out in a climate of apathy and economic restraints. In order to compete for limited expenditures and resources, the need for disaster countermeasures must be justified. In order to circumvent apathy, it is best to focus on predictable and likely events. The accomplishment of these objectives is facilitated by the collection of information about local hazards, the extent to which they threaten local populations, and the ease with which their effects can be averted.

The process of collecting this information is called hazard assessment or hazard analysis. At present, our ability to determine the probability of disaster events, the magnitude of their destructive potential, and the vulnerability of the populations, property, and natural resources they threaten is somewhat primitive and subjective (Kasperson, 1985:8; FEMA, 1984a:45,121). However, even a simplified hazard assessment based on subjective estimates gives a useful picture to guide planning priorities and to justify funding. These hazard assessment techniques are well described in several publications, including a number available from the Federal Emergency Management Agency, which can be consulted for further details (Lavalla, 1983:107; FEMA, 1984a:45; FEMA, 1984c:II-11; FEMA, 1983b:9; FEMA, 1983d:113,133,219; FEMA, 1985a; FEMA, 1985b; FEMA, 1985c).

Hazard Identification

For each hazard, provide the following information:

- Could this hazard affect your area?
- Is this hazard a *significant* threat in your area?

251

- About how often does this hazard pose a threat?
  Once a year or more?
  Once every 5 years?
  Once every 10 years?
  Once every 50 years?
  Once every 100 years?
  Less than every 100 years?
  Has not occurred?

- What is your best estimate of the total population that could be seriously affected by this hazard?

- What is the duration of impact to be expected from the hazard?

- What is the scope of damage to be expected from the hazard?

- What is the intensity of impact to be expected from the hazard?

- How predictable is the threat from this hazard?

- How easy is it to reduce the effects of the hazard?

Natural hazards:

Drought
Extreme cold
Extreme heat
Fire
   forest
   range
   other
Flood/flash flood
Avalanche
Landshift
   earthquake
   earthslide
   erosion
   subsidence (e.g., sink holes)
Snow/ice/hail
Dust/sand storms
Tsunami or storm surge
Volcanic eruption
Windstorm/tropical storm
Lightning storm
Hurricane
Tornado

Epidemic
  human
  animal
Blight/infestation (e.g., locust damage)

Technological hazards:

Hazardous materials accident
  fixed facility
  transportation
  chemical
  biological
  radioactive
Fire/explosion
Building/structure collapse
Dam/levee failure
Power/utility failure
Fuel shortage
Extreme air pollution (smog)
Transportation accident
  motor vehicle
  rail
  marine
  aircraft
  pipeline

Civil/political disorder:

Economic emergency
Riot
Strike
Demonstration
Terrorism/sabotage
Hostage incident
Enemy attack
  conventional
  biological warfare
  chemical warfare
  nuclear warfare

Secondary Hazards

In addition to describing the hazards that might threaten an area, it is also useful to list secondary threats that are likely to result from each of the above hazards (Lavalla, 1983:119).

Example: Hazards secondary to an earthquake:

Landslide
Building collapse
Hazardous material spill
Fire
Dam/levee failure
Interruption of communication/power/waste disposal/water supply/transpor-
    tation
Water pollution
Tsunami (tidal wave)
Seiche ("tidal wave" in an enclosed body of water, such as a lake)
Train wreck

Geographic Characteristics Affecting Vulnerability

What geographic and demographic features might affect your area's vulnerability to hazards?

Rivers
Canyons
Wildland areas
Earthquake vulnerable areas
    faults
    alluvial plains
    unstable hillsides
    mudflats
    landfill
Dams/levees
Power plants/switching stations
Water treatment facilities
Sewer lines and facilities
Mountains/hills
    avalanche hazard
    land/mud slide hazard
Flood plains
Coastal areas
Major highways
Rail lines/stations
Canals/rivers
Harbors
Airports

Pipelines
Power lines
Water storage areas
Mines
Manufacturing plants
Chemical/fuel storage areas
Toxic/radioactive dump sites
Military bases
Research labs
Prisons/detention centers/jails
Stadiums
Hospitals
Nursing homes
Senior citizen residences
Foreign language neighborhoods
Mobile home parks
Day care centers
Schools
Mental facilities
Emergency operations/dispatch/communications centers
Radio relay/satellite ground stations
Concentrations of crops and livestock

Hazard Maps

Hazard maps are useful tools for depicting the results of hazard assessments. A general map of the area, preferably topographical, and consistent with a standardized emergency operations mapping system, is created with clear acetate overlays. These overlays can be used to depict the threats and vulnerable structures/populations from particular types of hazards, such as those from earthquake, flooding, dam failure, and radioactive release from a fixed nuclear facility.

DISASTER RESOURCE-ACQUISITION DIRECTORY

The following is a listing of resources that might be useful in responding to disasters. It is recommended that this information be cross-tabulated in several ways, so that the information can be found quickly. Among the means by which the data can be organized are the following: alphabetically, by geographical area, by type of disaster, by agency, and by type of resource. If there are special procedures or contracts related to resource acquisition, they should also be specified. This information is most useful if computer-based with computer access available to multiple jurisdictions and agencies. However, if computers are used, they must have adequate data back-up capabilities and protection against electrical failure, "brown-out," electrical surge, and seismic activity or water exposure.

PERSONNEL AND ORGANIZATIONS

Personnel:
Alert Lists
Emergency Service and Civil Defense
   City
   County
   State
   Federal
Government Chief Executives
Health Officers
   City
   County
   State
Mutual Aid Coordinators
Regional Emergency Medical Services Agencies

Military:
Air Force Rescue Coordination Center
Local Military Bases and National Guard Units
U.S. Army Corps of Engineers

Law Enforcement Agencies:
Airport Police
Arson Investigators
City Police
Constables
County Sheriffs
Park Police
Forest Rangers
Fish and Game Wardens
University Police
State Police and Highway Patrol
Fire Marshals
FBI
U.S. Marshal
Secret Service
Bureau of Alcohol, Tobacco, and Firearms (ATF)
Environmental Protection Agency Agents
Coast Guard
Border Patrol
Treasury Agents (Customs)
Drug Enforcement Administration
Bomb Squads
SWAT Teams
Forensic or Body Identification Teams

Fire Suppression:
City Fire Departments
County and District Fire Departments
State Forestry
State Emergency Services Fire and Rescue Division
U.S. Forest Service
Bureau of Land Management
Airport Fire Departments
Private Manufacturing and Storage Facility Fire Brigades

Ambulance and Rescue Units:
Land Ambulances
Air Ambulances
   Fixed Wing
   Helicopter

Triage and Medical Teams

Heavy Rescue Units

Search and Rescue Teams:
Military Air and Sea Rescue
Civil Air Patrol
Wilderness (Mountain/Desert) Search and Rescue Teams
Mounted Posses
Cave Rescue Teams
Swift Water Rescue Teams
Underwater Rescue and Recovery Teams
4-Wheel Drive Search and Rescue Teams
Snowmobile Search and Rescue Teams
Cross-Country Ski Search and Rescue Teams
National Ski Patrol Units
Avalanche Rescue Teams
Search Dog Units
Avalanche Dog Units
Tracking Experts (e.g., Border Patrol)
Explorer Scout Search and Rescue Teams

Hazardous Material Response Teams:
Governmental
Private

Coroner or Body Identification Teams and Funeral Services

Animal Control, Humane Society and Veterinarians

Crisis Counseling Services

Suicide Prevention Services

Public Works and Highway Maintenance

Building Inspection Departments, Engineering Departments and Consultants

Utilities (Public and Private):
Natural Gas
Propane
Water
Electricity
Sewer
Telephone
Cable TV

Clergy

Foreign Language and Sign Language (Deaf) Translators

Governmental Welfare Agencies

Volunteer Welfare and Relief Organizations

Trade and Professional Associations, Unions and Service Clubs

Communications:
Amateur Radio Clubs
Citizens Band Clubs
News Media (Indicate if they have back-up power, satellite communication
    capability, mobile units, or foreign language correspondents)
    TV Stations
    Commercial Radio Stations
    Newspaper Offices
    News and Wire Services
    County or City Press Rooms
Two-Way Radio Service and Repair Services

Other Governmental Agencies:
Corrections Departments
Occupational Health and Safety
U.S. Geological Survey
State Mines and Geology
Seismic Safety Offices
Flood Control Districts (Dams and Levees)
U.S. Weather Bureau
Housing and Urban Development (HUD)
Small Business Administration
Nuclear Regulatory Commission
Department of Energy
U.S. Department of Transportation
    Federal Aviation Administration
    National Transportation Safety Administration
    Federal Railroad Administration
    Federal Maritime Administration
U.S. State Department
U.S. Bureau of Indian Affairs

FACILITIES

Medical:
Hospitals
    Trauma Centers

Burn Centers
Pediatric and Neonatal Intensive Care Services
Neurotrauma Centers
Spinal Cord Injury Treatment Centers
Reimplantation Centers
General Hospitals
Military and Veterans Hospitals
Public Health Hospitals
Sanitoriums
Hazardous Material Decontamination Facilities
Casualty Collection Points
Triage Centers
First Aid Centers
Portable Hospitals
Blood Banks
Drop-in Urgent Care Centers
Outpatient Surgical Facilities
Nursing Homes
Board and Care Facilities
Mental Health Facilities
   Inpatient
   Outpatient
Private Clinics and Medical Offices
Public Health Clinics
Medical Laboratories
Private X-Ray Offices
Animal Hospitals
Funeral Homes
Morgues
Refrigeration Lockers and Storage Rooms

Shelter and Feeding:
Red Cross and Salvation Army Shelters
Public Evacuation Shelters
Mobile Canteens
Catering Services
Portable Toilets
Schools and Colleges
Daycamps
Recreation and Community Centers
Meeting Halls and Recreation Centers
Roller Skating Rinks
Stadiums
Motels and Hotels

Campgrounds
Trailer Courts
Churches or Temples
Military Facilities
Portable Kitchens
Tents and Sporting Goods
Trailer Sales Offices
Restaurants
Banquet Halls
Parking Garages
Livestock and Animal Shelters or Facilities

Transportation Facilities:
Airports
Heliports
Marinas
Train Depots
Corporation Yards
Bus Stations
Ports and Harbors
Carports

TRANSPORTATION

(If radio-equipped, indicate frequencies and squelch tones)

Ambulance:
Land
Fixed Wing
Helicopter
4-Wheel Drive
Snowmobile
Marine
Physician or Nurse-staffed
Advanced Life Support (Paramedic)
Basic Life Support (Emergency Medical Technician)

Rescue:
Animal Control, Humane Society and Livestock Transport Vehicles
Light Rescue Units
Heavy Rescue Units
Fire Suppression Vehicles
Airport Crash Trucks
Buses and Taxis

Communications:
Mobile and Portable Communications Centers
Mobile Satellite Communications Units
Mobile Command Posts
Portable Radio Repeaters

Miscellaneous:
Mobile Canteens
Boats and Ships
Icecraft
Snow cats and Snowmobiles
Hovercraft
Amphibious Vehicles
Helicopters
Airplanes
  Passenger
  Cargo
  Tanker
  Fire Retardant Bomber
  Law Enforcement Pursuit
Rail transport
4-Wheel Drive
Barges
Tow Trucks and Heavy Duty Tow Trucks
Tank Trucks
  Fuel
  Water
  Milk
Refrigerator Trucks
Dump Trucks
Flat Bed Trucks
Trailers
  Cargo
  House
  Tank
  Livestock
Horses and Mules

EQUIPMENT AND SUPPLIES

Heavy Construction and Earthmoving Equipment:
Cranes
  Mobile

Rail
Barges
Bulldozers
Earth Drilling Equipment
Explosives
Snow Plows
Highway
Rail
Fork Lifts
Graders
Hoists
Loaders
Steam Shovels
Mixers
Rollers
Tractors

Construction:
Acetylene Torches
Arc Welding Equipment
Power Saws
Chain
Concrete
Wood
Plywood and Lumber
Hardware
Tarps and Plastic Sheeting
Fencing, Barricades and Traffic Cones
Sand or Salt
Sand Bags
Electrical Wire
Winches
Ladders
Pumps
Electrical Generators
Flood Lights
Battery Chargers (Automotive)
Chain
Jacks
Jack Hammers

Rescue and Medical:
Bandages and Sterile Dressings
Splints
Rubber Exam Gloves

IV Fluids and Administration Kits
Tape
Emergency Drugs
Suction Units
Backboards
Stretchers
Gurneys
Floodlights
Electrical Generators
Hydraulic Rescue Gear
   Hurst Tool
   Porto-Power
Air Bag Lifters
Cribbing
Come-Alongs
Self Contained Breathing Apparatus (SCBA)
SCBA Refilling Equipment
Decontamination Showers, Tubs and Whole-Body Basins
Hazmat Suits
Headlights, Flashlights, Batteries, Bulbs and Recharging Units
Rope Rescue Gear
Exposure Suits
SCUBA Gear and Wetsuits
Winches
Cervical Collars
Traction Splints
Vacuum Splints
Oxygen Supplies
Triage Tags and Forms
Tarps and Tents
Radiation Monitors
Hazardous Chemical Analysis Equipment
Hazardous Chemical References and Data Bases
Two-Way Radios, Batteries and Rechargers
Gas Detectors
Ladders
Infrared Detectors
Night Vision Equipment
Body Bags
Drag Lines
Maps
Portable Heating Equipment
Blankets and Sleeping Bags
Refrigerator Trucks

Office and Management Supplies:
Paper
Pens and Pencils
Clip Boards
Typewriters
Computers and Word Processors
Copying Equipment
Blackboards and Grease Pencil Boards
Desks
Staplers
Dictation Equipment, Tape Recorders and Video Equipment
Photography Equipment
Batteries
Lighting
Air Conditioning and Heating
Maps (Standardized)
  Street Maps
  Topographical (Contour)
  Hazard Maps
  Special Maps and Building Plans
    Shopping Centers
    Manufacturing Plants
    Fuel and Chemical Storage Sites
    Ports and Harbors
    Airports
    Stadiums
    Schools
    Hospitals and Medical Centers
    Mental Facilities
    Prisons, Detention Centers and Jails
    Large Public Buildings
    Shopping Centers
    Trailer Parks
    Parks
    Research Centers
    Power Plants (e.g., nuclear)
    Military Bases
    Marine
    Flood Plain Maps
    Evacuation Routes
    Pipeline Maps
      Water
      Fuel Chemical
      Sewer

Power Line Maps
Seismic Risk Maps
Watershed Maps
Orthophoto Maps

Shelter:
Water Storage Equipment
Portable Stoves and Fuel
Heating
Tents and Tarps
Tables
Cots, Cribs and Playpens
Sleeping Bags, Blankets, Sheets and Pillows
Portable Toilets
Portable Showers
Soaps and Towels
Clothing
Baby Supplies
Sanitary Napkins
Food and Utensils
Dishwashing Facilities
Trash Containers
Garbage and Plastic Bags
Bottled Water
Water Purification Supplies and Equipment

COMMUNICATIONS

Equipment and Supplies:
"Essential Service" Telephone Lines
Phone Booths
Walkie-Talkies (Indicate band and frequency)
Two-Way Radio Caches
Cellular Telephone
Satellite Communications Equipment
Signs, Barricades and Traffic Cones
Computer Terminals
Two-Way Radio Equipped Transportation (See under Transportation)
Portable Telephones and Intercoms
Ground-to-Air Panels
Aviation, Marine, Military and Business Band and CB radios
Mobile Communications and Command Posts
    Civilian
    Military

Scanning Receivers
Portable Televisions
Portable Commercial Broadcast Receivers
Weather Band Radios
Megaphones and Portable P.A. Systems
Maps

Telephone Number and Radio Frequency Listings:
Government
  Civilian
    (Refer to Headings of Organizations and Facilities)
  Military
Ambulances
Rescue Services
Search and Rescue Teams
Shelters
Casualty Collection Points
Staging Areas
Funeral Services
Animal Control and Humane Society
Mass Media (Commercial transmitters can be used for two-way transmission in
  disasters if other communications routes are unavailable.)
Cable TV
Mobile Telephone Numbers
Private Vehicles with Two-Way Radios
Taxis, Buses, Helicopters, Trains and Subways
Utilities
Relief and Welfare Agencies
Trucking and Towing Services
Delivery Services
Railroad
Airlines and Airports
Entertainment
  Raceways
  Amusement Parks
Private Manufacturing and Storage Facilities
Ham Repeaters
Private Security Services
Schools, School Boards, Colleges and Universities
Command Posts
Agency-Specific Mutual Aid Frequencies
Inter-Agency Coordination and Calling Frequencies
Frequency Sharing Agreements

Weather Bureau Frequencies
Microwave Numbers
Satellite Frequencies and Locations
Pay Phone Numbers
Emergency Broadcast System Procedures

Special Sources of Information:
Weather Bureau
Poison Control
CHEMTREC
Seismographic Stations
Flood Control Districts
Air Traffic Control
Geological Survey

INCIDENT COMMAND SYSTEM FORMS

PAGE 1

INCIDENT BRIEFING	1. INCIDENT NAME	2. DATE PREPARED	3. TIME PREPARED

4. MAP SKETCH

201	ICS 3-82	PAGE 1	8. PREPARED BY (NAME AND POSITION)

ICS Form 201

PAGE 2

7. SUMMARY OF CURRENT ACTIONS

201	ICS 3-82	PAGE 2	

ICS Form 201 (*continued*).

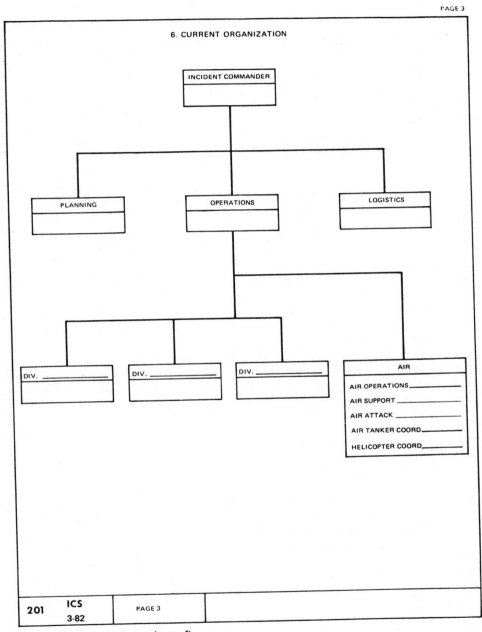

ICS Form 201 (*continued*).

PAGE 4

	5. RESOURCES SUMMARY			
RESOURCES ORDERED	RESOURCE IDENTIFICATION	ETA	ON SCENE ✔	LOCATION/ASSIGNMENT
201 ICS 3-82	PAGE 4			

ICS Form 201 (*continued*).

INCIDENT OBJECTIVES ICS 202	1. INCIDENT NAME	2. DATE PREPARED	3. TIME PREPARED

4. OPERATIONAL PERIOD (DATE/TIME)

5. GENERAL CONTROL OBJECTIVES FOR THE INCIDENT (INCLUDE ALTERNATIVES

6. WEATHER FORECAST FOR OPERATIONAL PERIOD

7. GENERAL/SAFETY MESSAGE

8. ATTACHMENTS (✓ IF ATTACHED)

☐ ORGANIZATION LIST (ICS 203)   ☐ MEDICAL PLAN (ICS 206)   ☐ _____

☐ DIVISION ASSIGNMENT LISTS (ICS 204)   ☐ INCIDENT MAP   ☐ _____

☐ COMMUNICATIONS PLAN (ICS 205)   ☐ TRAFFIC PLAN   ☐ _____

202 ICS 3-80	9. PREPARED BY (PLANNING SECTION CHIEF)	10. APPROVED BY (INCIDENT COMMANDER)

ICS Form 202.

ORGANIZATION ASSIGNMENT LIST ICS-203 1/82		1. INCIDENT NAME	2. DATE PREPARED	3. TIME PREPARED
POSITION	**NAME**	4. OPERATIONAL PERIOD (DATE/TIME)		
5. INCIDENT COMMANDER AND STAFF				
INCIDENT COMMANDER		9. OPERATIONS SECTION		
DEPUTY		CHIEF		
SAFETY OFFICER		DEPUTY		
INFORMATION OFFICER		a. BRANCH I — DIVISIONS/GROUPS		
LIAISON OFFICER		BRANCH DIRECTOR		
6. AGENCY REPRESENTATIVES		DEPUTY		
AGENCY	NAME	DIVISION/GROUP		
		DIVISION/GROUP		
		DIVISION/GROUP		
		DIVISION/GROUP		
		DIVISION/GROUP		
		b. BRANCH II — DIVISIONS/GROUPS		
		BRANCH DIRECTOR		
7. PLANNING SECTION		DEPUTY		
CHIEF		DIVISION/GROUP		
DEPUTY		DIVISION/GROUP		
RESOURCES UNIT		DIVISION/GROUP		
SITUATION UNIT		DIVISION/GROUP		
DOCUMENTATION UNIT		DIVISION/GROUP		
DEMOBILIZATION UNIT		c. BRANCH III — DIVISIONS/GROUPS		
TECHNICAL SPECIALISTS		BRANCH DIRECTOR		
		DEPUTY		
		DIVISION/GROUP		
		DIVISION/GROUP		
		DIVISION/GROUP		
		DIVISION/GROUP		
8. LOGISTICS SECTION		DIVISION/GROUP		
CHIEF		d. AIR OPERATIONS BRANCH		
DEPUTY		AIR OPERATIONS BR. DIR.		
a. SUPPORT BRANCH		AIR ATTACK SUPERVISOR		
DIRECTOR		AIR SUPPORT SUPERVISOR		
SUPPLY UNIT		HELICOPTER COORDINATOR		
FACILITIES UNIT		AIR TANKER COORDINATOR		
GROUND SUPPORT UNIT		10. FINANCE SECTION		
b. SERVICE BRANCH		CHIEF		
DIRECTOR		DEPUTY		
		TIME UNIT		
COMMUNICATIONS UNIT		PROCUREMENT UNIT		
MEDICAL UNIT		COMPENSATION/CLAIMS UNIT		
FOOD UNIT		COST UNIT		
203 ICS 1/82	PREPARED BY (RESOURCES UNIT)			

ICS Form 203.

1. BRANCH	2. DIVISION/GROUP	**DIVISION ASSIGNMENT LIST**	**ICS 204** (1-82)

3. INCIDENT NAME	4. OPERATIONAL PERIOD
	DATE _____
	TIME _____

5. OPERATIONS PERSONNEL

OPERATIONS CHIEF _____ DIVISION/GROUP SUPERVISOR _____

BRANCH DIRECTOR _____ AIR ATTACK SUPERVISOR _____

6. RESOURCES ASSIGNED THIS PERIOD

STRIKE TEAM/TASK FORCE/ RESOURCE DESIGNATOR	LEADER	NUMBER PERSONS	TRANS. NEEDED	DROP OFF PT./TIME	PICK UP PT./TIME

7. CONTROL OPERATIONS

8. SPECIAL INSTRUCTIONS

9. DIVISION/GROUP COMMUNICATION SUMMARY

FUNCTION		FREQ.	SYSTEM	CHAN.	FUNCTION		FREQ.	SYSTEM	CHAN.
COMMAND	LOCAL				SUPPORT	LOCAL			
	REPEAT					REPEAT			
DIV./GROUP TACTICAL					GROUND TO AIR				

PREPARED BY (RESOURCE UNIT LDR.)	APPROVED BY (PLANNING SECT. CH.)	DATE	TIME

ICS Form 204.

INCIDENT RADIO COMMUNICATIONS PLAN		1. INCIDENT NAME	2. DATE/TIME PREPARED	3. OPERATIONAL PERIOD DATE/TIME

4. BASIC RADIO CHANNEL UTILIZATION

SYSTEM/CACHE	CHANNEL	FUNCTION	FREQUENCY	ASSIGNMENT	REMARKS

5. PREPARED BY (COMMUNICATIONS UNIT)

ICS 8-78

205

ICS Form 205.

MEDICAL PLAN	1. INCIDENT NAME		2 DATE PREPARED	3. TIME PREPARED	4 OPERATIONAL PERIOD	

5. INCIDENT MEDICAL AID STATIONS

MEDICAL AID STATIONS	LOCATION	PARAMEDICS	
		YES	NO

6. TRANSPORTATION

A. AMBULANCE SERVICES

NAME	ADDRESS	PHONE	PARAMEDICS	
			YES	NO

B. INCIDENT AMBULANCES

NAME	LOCATION	PARAMEDICS	
		YES	NO

7. HOSPITALS

NAME	ADDRESS	TRAVEL TIME		PHONE	HELIPAD		BURN CENTER	
		AIR	GRND		YES	NO	YES	NO

8. MEDICAL EMERGENCY PROCEDURES

9. PREPARED BY (MEDICAL UNIT LEADER)	10. REVIEWED BY (SAFETY OFFICER)

206 **ICS** 8-78

ICS Form 206.

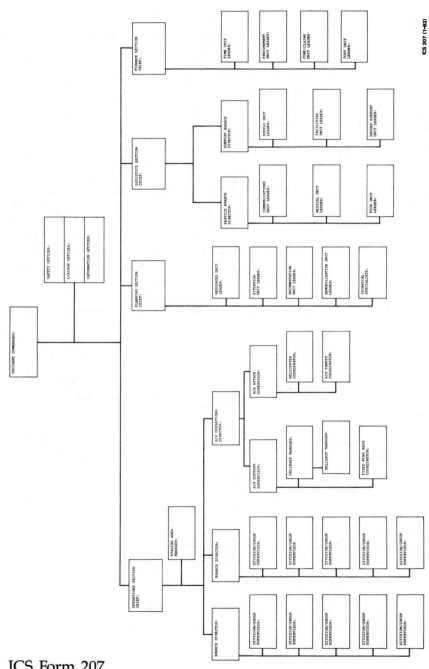

ICS 207 (1-82)

ICS Form 207.

1. INCIDENT NAME		2. INCIDENT NO.	3. INCIDENT COMMANDER		4. JURIS-DICTION	5. COUNTY	**INCIDENT STATUS SUMMARY** ICS 209 (1-81)	
6. TYPE INCIDENT	7. LOCATION						8. STARTED (DATE/TIME)	
9. CAUSE		10. AREA INVOLVED	11. PERCENT CONTAINED	12. EXPECTED CONTAIN-MENT Date___ Time___		13. PERCENT CONTROLLED	14. EXPECTED CONTROL Date___ Time___	
15. CURRENT THREAT				16. CONTROL PROBLEMS				
17. EST. LOSS	18. EST. SAVINGS		19. INJURIES___ DEATHS___	20. LINE BUILT		21. LINE TO BUILD		
22. CURRENT WEATHER WS WD	TEMP RH		23. PREDICTED WEATHER NEXT PERIOD WS WD	TEMP RH		24. INCIDENT COSTS—PREVIOUS DAY	25. TOTAL COST TO DATE	

26. AGENCIES																									TOTALS	
27. RESOURCES KIND OF RESOURCE	INC	ST	INC	ST	INC	ST	INC	ST	INC	ST	INC	ST	INC	ST	INC	ST	INC	ST	INC	ST	INC	ST	INC	ST	INC	ST
ENGINES																										
DOZERS																										
CREWS																										
HELICOPTERS																										
AIR TANKERS																										
TRUCK COS.																										
RESCUE/MED.																										
WATER TENDERS																										
OVERHEAD PERSONNEL																										
TOTAL PERSONNEL																										

28. COOPERATING AGENCIES

29. REMARKS

30. PREPARED BY	31. APPROVED BY	32 DATE ___ TIME ___	33. INITIAL ☐ UPDATE ☐ FINAL ☐	34. SENT TO DATE ___ TIME ___ BY ___

7540-130-0289

ICS Form 209.

CHECK-IN LIST ICS-211 (1#2)

1. INCIDENT NAME

2. CHECK-IN LOCATION   ☐ BASE   ☐ CAMP   ☐ STAGING AREA   ☐ ICP RESTAT   ☐ HELIBASE

3. DATE/TIME

4. LIST PERSONNEL (OVERHEAD) BY AGENCY & NAME
   - OR -
   LIST EQUIPMENT BY THE FOLLOWING FORMAT:

CHECK-IN INFORMATION

4. AGENCY	SINGLE / T/F S/T KIND	TYPE	I.D. NO. / NAME	5. ORDER/ REQUEST NUMBER	6. DATE/TIME CHECK-IN	7. LEADER'S NAME	8. TOTAL NO. PERSONNEL	9. MANIFEST YES NO	10. CREW WEIGHT OR INDIVIDUALS WEIGHT	11. HOME BASE	12. DEPARTURE POINT	13. METHOD OF TRAVEL	14. INCIDENT ASSIGNMENT	15. OTHER QUALIFICATION	16. SENT TO RESTAT TIME/INT. STAT TIME/INT.

17. Page _____ of _____

18. PREPARED BY (NAME AND POSITION)   USE BACK FOR REMARKS OR COMMENTS

ICS Form 211.

UNIT LOG ICS-214		1. INCIDENT NAME	2. DATE PREPARED	3. TIME PREPARED
4. UNIT NAME/DESIGNATOR		5. UNIT LEADER (NAME AND POSITION)	6. OPERATIONAL PERIOD	

7.	PERSONNEL ROSTER ASSIGNED	
NAME	ICS POSITION	HOME BASE

8.	ACTIVITY LOG (CONTINUE ON REVERSE)
TIME	MAJOR EVENTS

ICS Form 214.

TIME	MAJOR EVENTS

| 214 | ICS 5-80 | 9. PREPARED BY (NAME AND POSITION) |

ICS Form 214 (*continued*).

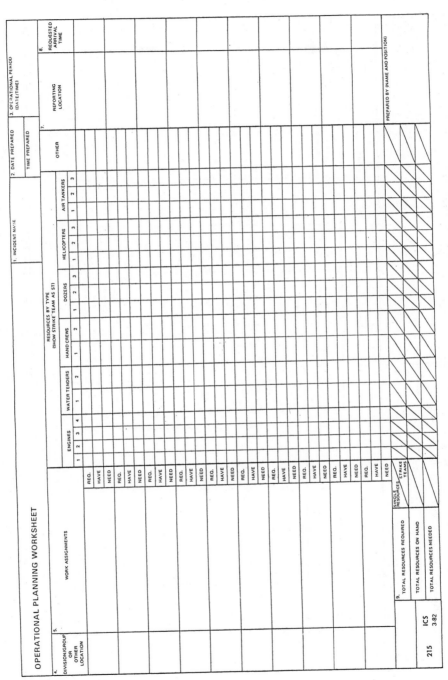

ICS Form 215.

RADIO REQUIREMENTS WORKSHEET

1. INCIDENT NAME
2. DATE
3. TIME
4. BRANCH
5. AGENCY
6. OPERATIONAL PERIOD
7. TACTICAL FREQUENCY
8. DIVISION/GROUP
AGENCY
9. AGENCY

ID NO. | RADIO RQMTS | AGENCY

DIVISION/GROUP
AGENCY

ID NO | RADIO RQMTS | AGENCY

DIVISION/GROUP
AGENCY

ID NO | RADIO RQMTS | AGENCY

DIVISION/GROUP
AGENCY

ID NO | RADIO RQMTS

10. PREPARED BY (NAME AND POSITION)

PAGE

216

ICS 3-82

ICS Form 216.

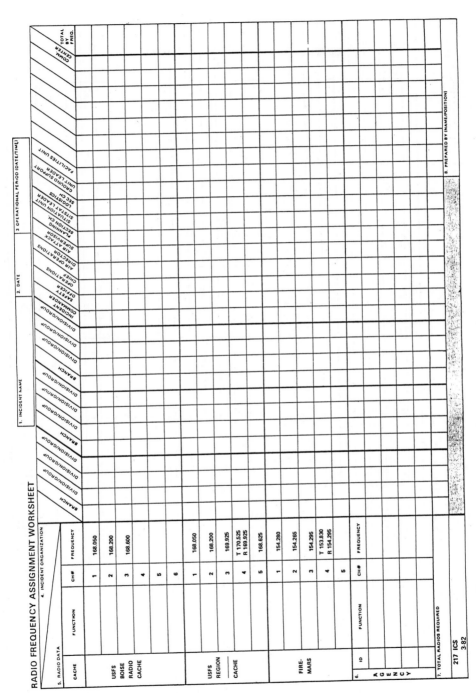

ICS Form 217.

SUPPORT VEHICLE INVENTORY
(USE SEPARATE SHEET FOR EACH VEHICLE CATEGORY)

1. INCIDENT NAME	2. DATE PREPARED	3. TIME PREPARED

4.

VEHICLE INFORMATION					
a. TYPE	b. MAKE	c. CAPACITY/ SIZE	d. AGENCY/OWNER	e. I.D. NO.	f. LOCATION	g. RELEASE TIME

5. PREPARED BY (GROUND SUPPORT UNIT)	
PAGE	

218   ICS
      8-78

ICS Form 218.

AIR OPERATIONS SUMMARY

1. INCIDENT NAME		2. OPERATIONAL PERIOD		3. DISTRIBUTION	
		DATE	TIME	HELIBASES	FIXED WING BASES

PERSONNEL & COMMUNICATIONS	NAME	AIR/AIR FREQUENCY	AIR/GROUND FREQUENCY	5. REMARKS (Specific Instructions, Safety Notes, Hazards, Priorities)
AIR OPERATIONS DIRECTOR				
AIR ATTACK SUPERVISOR				
HELICOPTER COORDINATOR				
AIR TANKER COORDINATOR				

6. LOCATION/ FUNCTION	7. ASSIGNMENT	8. FIXED WING		9. HELICOPTERS		10. TIME		AIRCRAFT ASSIGNED	OPERATING BASE
		NO.	TYPE	NO.	TYPE	AVAILABLE	COMMENCE		
13. TOTALS									

14. AIR OPERATIONS SUPPORT EQUIPMENT	15. PREPARED BY	DATE	TIME

ICS 220 (4/86)

ICS Form 220.

5-CATEGORY TRIAGE CLASSIFICATION EXAMPLES

NOTE: This list gives examples of conditions that might be found in disaster casualties and how they would be classified in the 5-category system described in Chapter 8. Since it is not always possible to make the correct diagnosis in the field, important signs and symptoms are also listed. The reader may notice that the examples include conditions more reminiscent of routine emergency medical conditions rather than those typical of disasters. They are included for two reasons:

1) People do not cease becoming ill or having babies simply because disaster strikes. Those with non-disaster related medical conditions will still require treatment and need to be considered when priorities are assigned.

2) If triage categories are to be used on a daily basis, then one must be able to determine triage categories for routine emergencies.

EXAMPLES OF "CRITICAL" (RED) CASUALTIES

- Upper airway obstruction.

- Life-threatening bleeding.

- Tension pneumothorax.

- Contamination with a hazardous substance.

- Second- or third-degree facial burns (when the total body surface area burned is less than 50%)—which may rapidly lead to upper airway obstruction.

- Stridor (crowing or raspy inspiratory sounds)—suggests upper airway obstruction.

- Severe, rapidly progressing allergic symptoms such as: rash, generalized or facial swelling, wheezing, stridor or breathing difficulty, and weak, thready pulse.

- Severe sore throat when accompanied by drooling, muffled voice, inability to swallow, or difficulty opening the jaw—suggests a serious infection in the airway which may quickly lead to its obstruction.

- Complicated obstetrical delivery (e.g., breech position, or compressed umbilical cord).

- Cardiac rhythm abnormality if accompanied by the sudden onset of circulatory shock, decreased mental alertness, or pain, burning, pressure or tightness in the chest, upper abdomen, upper extremity, neck, jaw, or back—suggests a heart attack.

- Untreated poisoning (after initial treatment such as the administration of syrup of ipecac, some cases may be retriaged to a lower priority category).

- Hypoglycemia, insulin shock, or insulin reaction (severe decrease in blood sugar which may occur in diabetics).

- Respiratory distress (blue skin color, asymmetrical chest motion or sounds, noisy breathing, nasal flaring, tightening of the neck muscles during breathing efforts, or retraction of the skin between the ribs, about the collar bones or above or below the breastbone).

- Circulatory shock.

- Rapidly deteriorating level of consciousness.

- Rapidly progressive nervous system disorder (paralysis, weakness, numbness, tingling, incoordination, confusion, or visual disturbance).

- Seizure during pregnancy (suggests eclampsia or toxemia).

- Status epilepticus (more than two seizures without regaining full consciousness in between).

- Penetrating wounds of the chest, abdomen, pelvis, rectum, vagina, head or neck.

- Embedded radioactive foreign bodies.

- Sunstroke or fever greater than 105° F.

- Coma.
- Untreated cervical spine injuries (after these are properly immobilized, their priority for care may decrease).

EXAMPLES OF "CATASTROPHIC" (BLUE) CASUALTIES

- Cardiac arrest (especially resulting from trauma or blood loss).
- Respiratory arrest when not due to drugs or upper airway obstruction.
- Sudden onset of severe abdominal pain and circulatory shock in an elderly subject who has a pulsating abdominal mass—suggests a ruptured abdominal aneurysm.
- Massive brain injuries (indicated by massive head trauma, dilated-fixed pupils, absence of all reflexes, extrusion of brain matter).
- Second and/or third degree burns exceeding a total of 50% of the body surface (especially in elderly persons with serious underlying medical disorders such as emphysema, diabetes, or cirrhosis of the liver).
- Penetrating wounds of the trunk with coma and no detectable blood pressure.
- Penetrating gunshot wounds of the head with coma.
- Cardiogenic shock.

EXAMPLES OF "URGENT" (YELLOW) CASUALTIES

- Circulatory shock which has responded adequately to initial treatment with one liter of IV fluid.
- Interference with circulation due to a fracture or dislocation.
- Severe bleeding controlled by a tourniquet.
- Compartment syndrome (swelling due to an injury, usually a fractured elbow or shin, which interferes with capillary blood flow to the muscle tissue; usually characterized by severe pain which is aggravated by movement of the joints beyond the injury and, *sometimes*, by a sensory or pulse deficit).

- Dislocations of the hip, elbow, or knee—which may compress arteries or nerves, or may be difficult to reduce if treatment is delayed.

- Open dislocations and fractures.

- Acute urinary retention (especially in chronically paralyzed patients).

- Second- or third-degree burns (not including the face or airway, and totaling less than 50% of the body surface).

- Uncomplicated bends (decompression sickness, caisson's disease).

- Apparently complete recovery after resuscitation from drowning.

- Electrical burns.

- Non-severe bleeding from the genitalia, digestive tract or lungs in the absence of circulatory shock.

- Uncomplicated obstetrical delivery or impending delivery (some might even consider this belonging to category Green).

- Severe headache not related to an injury, with decreased alertness, or with confusion, fever, or a stiff neck (inability of the patient to touch his chin to his chest)—suggests meningitis or an infection or bleeding in the brain.

- Uncomplicated hypothermia (rectal temperature less than 95° F).

- Swelling of a digit or the genitalia due to a constricting object or band.

- Sustained blood pressure exceeding 200 mm systolic or 120 mm diastolic, especially in pregnancy.

- Severe abdominal pain with abdominal wall rigidity or localized tenderness—suggests internal bleeding or infection such as that due to a perforated stomach ulcer or a ruptured appendix.

- Pelvic fractures in the absence of circulatory shock.

- Smoke inhalation in the absence of respiratory distress.

- Uncomplicated gunshot or stab wound to an extremity.

- Multiple fractures in the absence of shock.

- Immobilized, uncomplicated cervical spine injuries.

- Traumatic crush injuries or amputation of an extremity in the absence of serious bleeding or circulatory shock (except crush injuries of a finger tip, which may be in category Green).

- Fever with severe joint pain, or fever in a child or infant who refuses to use an extremity—suggests a severe infection of a joint.

- Sudden onset of confusion, disorientation, combativeness, or psychotic behavior (when not due to injury, hypoglycemia, poisoning or overdose, shock, or oxygen deficiency).

- First onset of severe, incapacitating headache—suggests meningitis, brain infection, or bleeding in the brain.

- Sudden onset, but not rapidly progressive, localized sensory loss or abnormality, partial or complete paralysis, or sustained loss of balance.

- Sudden, partial or complete, temporary or sustained, but not progressively deteriorating abnormality of vision.

- Penetrating wound of the eyeball.

- Rectal temperature greater than 104° F in a child.

- Fever in a child who is unusually lethargic or refuses to eat or play.

- Rectal temperature greater than 100° F in an infant less than 3 months old.

- Oral temperature greater than 103° F in an adult.

- Uncomplicated femoral (thigh) fractures.

- Vaginal bleeding in pregnancy.

- Vaginal bleeding accompanied by light-headedness, fainting, or severe back, abdominal, or shoulder pain.

- Sudden onset of severe testicular pain—suggests a twisting of the testicular cord which has cut off the blood supply.

- Soft tissue infections of the face, especially about the eye or nose—can travel to the brain if not treated.

- Soft tissue infections or animal (especially human) bites of the hand can cause, within hours, permanent damage to the hand.

- Wounds of the external genitalia (in the absence of shock or continued bleeding).

- First onset of seizures (in the absence of status epilepticus, hypoglycemia, poisoning or overdose, injury, cardiac rhythm disturbance, or oxygen deficiency).

- Vomiting more than twice after a head injury.

- Repeated vomiting or diarrhea in a child who is abnormally lethargic, has a weak cry, dry tongue or inability to make tears—suggests serious illness or dehydration.

- Repeated vomiting in a diabetic—suggests diabetes is out of control.

- Serious surface injury to the face or head (in the absence of uncontrolled bleeding, airway problems, circulatory shock, decreased mental alertness, confusion, vomiting or non-immobilized cervical spine injury).

- Large, dirty, or crushed soft tissue wounds.

EXAMPLES OF "MINOR" (GREEN) CASUALTIES

- Closed dislocations of the jaw, kneecap, or finger (especially if they have been reduced).

- Closed, uncomplicated fractures of the upper extremity, lower leg, foot, kneecap, ankle, or face.

- Uncomplicated, clean lacerations (including those involving tendons or peripheral nerves).

- Fingertip amputations with loss or crushing of the amputated part (which precludes sewing it back on).

- Burns in adults totaling less than 20% of the body surface area (when they do not involve the face, airway, groin or anal area, eye, feet, or hands).

- First degree burns not affecting the airway or eye.

- Frostbite.

- Sprains, strains, and moderate bruises.

- Dental pain in the absence of facial infection.

- Psychiatric or emotional disorders (when not due to physical injury or disorder and not involving suicidal or homicidal tendencies).

- Uncomplicated abrasions.

- Nosebleeds that can be stopped by direct pressure (firm pinching of the soft part of the nose).

SOURCES OF PUBLIC EDUCATION MATERIAL

The following is a list of references for material that can be used to educate the public or members of the press about various types of disasters, how to prepare for them, and what to do after impact. Also included are references on how to carry out public education programs and on model programs in existence.

ABC Media Concepts: Earthquake Parts I & II, movie, Xerox Films, Inc, 245 Long Hill Rd, Middletown, CT 06457.

American Educational Films, Tomorrow's Quake, 20 min. movie, 132 Lasky Dr, Beverly Hills, CA 90212, (213) 278-4996, $275.

American National Red Cross: Free Community Education Resource Guide, Emergency And Community Services, American National Red Cross, 1986, 122 pp. Obtainable from local Red Cross chapters.

American Red Cross, Los Angeles Chapter, Golden Gate Chapter, 1550 Sutter St, San Francisco, CA 94109, (415) 776-1500:
    Disaster Preparedness for Disabled and Elderly People, 1985.
    Safety and Survival in Earthquake, 1984.

American Red Cross, San Bernardino County Chapter, Family Disaster Plan and Personal Survival Guide, 1524 North E St, San Bernardino, CA 92402, (714) 888-1481, no date.

Alameda County Health Care Services Agency, Earthquake Country: A Household Guide to Earthquake Safety, (17 pg pamphlet). Source: Affirmative Action Officer, Alameda County Health Care Services Agency, 499 Fifth St, Oakland, CA 94607, (415) 874-7636, no date.

Area Disaster Council, The Family Earthquake Plan, San Mateo Area Office of Emergency Services, 401 Marshall St, Redwood City, CA 94063, (415) 363-4790, no date.

Bay Area Regional Earthquake Preparedness Project, MetroCenter, 101 Eighth St, Suite 152, Oakland, CA 94607, (415) 540-2713:
    Earthquake Preparedness and Public Information: An Annotated Bibliography, 1985.
    Marketing Earthquake Preparedness: Community Campaigns That Get Results, 1985.

Bolt BA: Earthquakes, A Primer, WH Freeman & Co, New York, 1978.

Boraiko AA: Storing Up Trouble . . . Hazardous Waste, National Geographic, 167(3): 319-351, March 1985.

Children's Television Workshop, Dept. CES/ARC, One Lincoln Plaza, New York, NY 10023, (800) 624-4800, (212) 595-3456:
The Big Bird Get Ready for Hurricanes Kit:
    Four-color booklet with safety tips
        A Hurricane Blues soundsheet
        A Hurricane Force board game with science facts and safety information
    The Big Bird Get Ready for Earthquake Kit
    Sesame Street Fire Safety Resource Book

EQE, Inc.: The EQE Earthquake Home Preparedness Guide, 121 Second St, San Francisco, CA, (415) 495-5500, 1985.

Federal Emergency Management Agency, FEMA, PO Box 70274, Washington, DC, 20024 (Catalog available on request):

In Time of Emergency: A Citizen's Handbook, Publication H-14/October, 1985.

Hurricane Awareness Workbook, Publication FEMA-86, 1985.

Emergency Management U.S.A., Publication SS-2/October, 1981.

Hurricanes: Safety Tips for Hurricanes, Publication L-105, 1980.

Tornado Safety Resource Workbook, 1983.

Preparing for Hurricanes and Coastal Flooding: A Handbook for Local Officials, Publication FEMA 50/October, 1983.

Tornado Preparedness Campaign Kit

Safety Tips for Winter Storms, L-96 (8/85) (Leaflet)

Winter—Fire Safety Tips in the Home, L-97, (10/87) (Leaflet)

Earthquakes, L-111 (7/83) (Leaflet)

After the Fire, L-114 (12/84) (Leaflet)

Learning to Live in Earthquake Country—Preparedness in Apartments and Mobile Homes, L-143 (9/86) (Leaflet)

Flash Floods, L-146 (9/85) (Leaflet)

Tornado Safety Tips, L-148 (2/86) (Leaflet)

Disaster Housing Assistance, L-151 (4/87) (Leaflet)

Dam Safety: Know the Potential Hazard, L-152 (5/87) (Leaflet)

Retrofitting Flood Prone Residential Structures, L-153 (6/87) (Leaflet)

Emergency Preparedness Checklist, L-154 (9/87) (Leaflet)

Winter Survival Coloring Book, FEMA-26 (7/85)

Earthquake Safety Checklist, FEMA-46 (10/85)

You Can Survive a Tornado: Safety Tips, FEMA-56 (4/84)

Earthquake Public Information Materials: An Annotated Bibliography, FEMA-67 (9/86)

When Disaster Strikes: A Handbook for the Media, FEMA-79 (9/85)

Guidebook for Developing a School Earthquake Safety Program, FEMA-88 (12/85)

Reducing the Risks of Nonstructural Earthquake Damage: A Practical Guide, FEMA-74, (6/85), Eratta Sheet (12/85)

Preparedness for People with Disabilities (Earthquakes), FEMA-75 (6/85)

Preparedness in High-Rise Buildings (Earthquakes), FEMA-76 (6/85)

Guidelines for Local Small Businesses in Meeting the Earthquake Threat, FEMA-87 (9/85)

A Guide to Marketing Earthquake Preparedness, FEMA-111 (9/86)

Marketing Earthquake Preparedness, FEMA-112 (9/86)

Family Earthquake Safety Home Hazard Hunt and Drill, FEMA-113 (9/86)

Survival in a Hurricane (Wallet Card, English/Spanish)

Tips for Tornado Safety (Wallet Card, English/Spanish)

Flash Flood (Wallet Card, English/Spanish)

Big Bird Get Ready for Hurricane Kit, K-61 (3/84)

Hurricane, Poster-2 (4/82)

Winter Watch for Kids, Poster-5 (10/82)

Earthquake, Poster-6 (9/83)

Hurricane Awareness: Action Guidelines for School Children

Hurricane Awareness:

Action Guidelines for Senior Citizens

Tornadoes and Severe Storms Awareness Campaign Workbook

Winter Safety Workbook

Winter Survival Test

Storm Surge and Hurricane Safety: With North Atlantic Tracking Chart, L-140 (6/84)

Perspectives on Hurricane Preparedness: Techniques in Use Today, (10/84)

Film Communicators: Earthquake Don'ts and Do's, 11 min. movie, 11136 Weddington St, North Hollywood, CA 91601, (213) 766-3747.

Foraker HW: What You Should Know About Earthquakes: It Could Save Your Life: An Earthquake Awareness Book, SJB Publishing Co, Mission Viejo, CA, 1983.

Funk B: Hurricane, National Geographic 158(3):346, September 1980.

Gere JM, and Haresh CS: Terra Non Firma: Understanding and Preparing for Earthquakes, Stanford Alumni Association, Stanford, CA, 1984.

Getting Ready for a Big Quake, Sunset Magazine, Lane Publishing, Menlo Park, CA, March 1982.

Groot I: School Earthquake Handbook, PO Box 20093, San Jose, CA 95160, 1984.

Hatfield S: What to do When a Tornado Strikes, Consumer's Research 67:18-20, April 1984.

Health Plus, Getting Ready for the Big One, 2966 Diamond St, Suite 434, San Francisco, CA 94131, (415) 585-2221, no date.

Iacopi R: Earthquake Country, Lane Books, Menlo Park, CA, 1976.

Kansas Division of Emergency Preparedness: Fire Awareness, 25 pp, PO Box C-300, Topeka, KS 66601.

Kessler E: Thunderstorms and Hurricanes, McGraw Hill Yearbook of Science and Technology, McGraw Hill Book Co, 1977.

Kingsford ST: The California Earthquake Manual, Kingsford Publishing, 42 Birchwood Ave, Agoura, CA, 1976.

Knotts D: Storm Warning! How to Weather a Hurricane, Survive 4:38, March 1984.

Kockelman WJ: Reducing Losses from Earthquakes Through Personal Preparedness, Open-File Report 84-765, Open File Services Center, USGS, PO Box 25425, Federal Center, Denver, CO 80225, 1984.

Lafferty L: Earthquake Preparedness: A Handbook for Home, Family, & Community, Lafferty & Assoc, Inc, PO Box 1026, La Cañada, CA 91011, (818) 952-5483, 1984.

Lavalla P: Handbook, Living Life's Emergencies: A Guide for Home Preparedness, Emergency Response Institute, 1919 Mark St, NE, Olympia, WA 98506, (206) 491-7785, (509) 782-4832, 1981.

Miller P: Tornado!, National Geographic 171(6):690-715, June 1987.

National Association for Search & Rescue: Emergency School Plan, PO Box 2123, La Jolla, CA 92038, 1980.

National Fire Protection Association, Batterymarch Park, Quincy, MA 02269, (800) 344-3555, Catalog available for:
How to Survive a Fire, in: Hotel Fires Behind the Headlines: A Compilation of Articles from Fire Command, Fire Journal, and Fire Technology, p 86, NFPA Publication no SPP-74.
Learn Not to Burn Curriculum,
Level 1: Grades K-2

Level 2: Grades 3-5
Level 3: Grades 6-8
Fire Prevention All Over Your Home (Brochure)
10 Tips for Firesafety (Brochure)
Exit: Escape From Fire Wherever You Are (Brochure)
Fire in the Kitchen (Flyers)
Firepower (Film)
Sparky's Coloring Book
I am Fire (Video)
Learn Not to Burn (Video/Film)
Learn Not to Burn Your Whole Life Through (Brochure)
Learn Not to Burn Wherever You Are (Video/Film)
Babysitter's Handbook
Fire Sleuths (Video/Film)
Project Burn Protection (Multi-media)
Fire Response (Video/Film)
An Ounce of Prevention (Video/Film)
Understanding Home Fire Detectors (Brochure)
Play it Safe . . . Plan Your Escape (Brochure)
What Do I Do When I See a Fire? (Video/Film)
Exit Drills in the Home (Video/Film)
Sparky Reports a Fire (Comic Book)
If You Get Caught in Smoke: Crawl (Flyer)
A Guide to Hurricane Preparedness Planning for State and Local Officials, CPG 2-16 (12/84)

National Oceanic and Atmospheric Administration, Superintendent of Documents, U.S. Government Printing Office, Washington, DC 20402:
Thunderstorms and Lightning, NOAA/PA 83000
Tornado Safety—Surviving Nature's Most Violent Storms, NOAA/PA 82001
Tips for Tornado Safety, NOAA/PA 76016
Tornado Safety in Residences, NOAA/PA 79016
In Your Keeping (RE Administrators of schools, Hospitals, etc.), NOAA/PA 70015
Tornado Safety Rules in Schools, NOAA/PA 74025
Tornado, Do You Know What To Do?, NOAA/PA 75015
Owlie Skywarn's Tornado Warning, NOAA/PA 75012
Skyward, Seconds Save Lives, NOAA/PA 70010
Spotter's Guide for Identifying and Reporting

Severe Local Storms, NOAA/PA 81011
Severe Local Storm Spotter Reporting Procedures Poster, NOAA/PA 70012
Skywarn Poster, NOAA/PA 76019
Tornado Emergency, Effective Procedures for All Broadcast Stations, NOAA/PA 78021
Skywarn Spotter, NOAA/PA 84001
Tornado Safety Rules Poster, Poster 923
Flash Floods, NOAA/PA 77014 (Spanish version NOAA/PA 77015)
Flash Floods, NOAA/PA 73018 (Spanish—NOAA/PA 74022)
Floods, Flash Floods and Warnings, NOAA/PA 81010
Killer from the Hills Poster, (Flash flood safety rules) NOAA/PA 73019 (Spanish—NOAA/PA 74021)
Owlie Skywarn on Flash Floods Poster, NOAA/PA 77016
Some Devastating North Atlantic Hurricanes of the 20th Century, NOAA/PA 77019
Storm Surge and Hurricane Safety with Tracking Chart, NOAA/PA 78019
Hawaiian Hurricanes and Safety Measures with Central Pacific Tracking Chart, NOAA/PA 85002
Owlie Skywarn's Hurricane Warnings, NOAA/PA 77001
Survival in a Hurricane, NOAA/PA 70027
Tsunami Watch and Warning, NOAA/PA 74301
Dust Storm, NOAA/PA 82002
Heat Wave A Major Summer Killer, NOAA/PA 85001
  Watch Out Storms Ahead: Owlie Skywarn's Weather Book, NOAA/PA 82004.

National Oceanic and Atmospheric Administration, Film and Slide Series, Order Section, National Audiovisual Center, General Services Administration, Washington, DC 20409, (301) 763-1896. Films are available on loan, free except return postage, from Modern Talking Pictures, Film Scheduling Center 5000 Park Street North, St Petersburg, FL 33709, (813) 541-5763:
  Thunderstorm (lightning) Slide Lectures
  Day of the Killer Tornadoes-15 min movie
  Tornado at Pleasant Hill-5 min movie
  Tornado: A Spotter's Guide-15½ min movie
  Terrible Tuesday (tornadoes)-23 min movie
  Tornado Preparedness-67 slides
  Tornado Safety in Residence-130 slides
  The Safest Place in Schools (tornadoes)-140 slides
  A Look at the Tornado and Other Local Storms-63 slides
  Flash Flood-14 min movie
  Flash Floods: Myths or Realities-79 slides
  Flash Flood Preparedness-80 slides
  Survival in the Cold-16 min movie
  Winter Storms-28 min movie
  Winter Storms, the Deceptive Killers-80 slides

National Weather Service: Winter Weather Information Packet, James R. Poirier, Warnings and Preparedness Meteorologist, National Weather Service Forecast Office, 7777 Walnut Grove, OMI, Memphis, TN 38119-2198.

Office of Emergency Services & California Earthquake Education Project: Elementary School Preparedness Kit, OES, PO Box 9577, Sacramento, CA 95823, 1985.

Office of the United Nations, Disaster Relief Coordinator, Disaster Prevention and Mitigation: A compendium of Current Knowledge, Vol 11: Preparedness Aspects, pp 87–101, Director, Liaison Office, United Nations Disaster Relief Office, New York, NY 10017, 1984.

Rubin CB: Disseminating Disaster-Related Information to Public and Private Users, Working Paper #47, Natural Hazards Research and Applications Information Center, IBS #6, Campus Box 482, University of Colorado, Boulder, CO 80309, (303) 492-6818, 1982.

San Bernardino County Office of Public Safety, A Concept for Disaster Preparedness, County of San Bernardino, Office of Public Safety, 1776 Miro Way, Rialto, CA 92376.

Scientific American, Earthquakes and Volcanoes, ed 10, WH Freeman & Co, San Francisco, 1980.

Snow JT: The Tornado, Scientific American 251:86-94, April 1984.

Southern California Earthquake Preparedness Project, American Red Cross, Los Angeles Chapter, Southern California Div, 2700 Wilshire Blvd, Los Angeles, CA 90057, (213) 739-5200:
Earthquake Public Information and Education Program Design, 1983.
Earthquake Preparedness Information for People with Disabilities, 1983.

Guidelines for School Earthquake Safety Planning, 1982.

SUNY College, Everyday Weather Project, Brockport, NY 14420, (716) 395-2352:
Hazardous Weather: Hurricanes (video)
Filmstrip/slide sets:
Sensing and Analyzing Weather
Weather Systems
Weather Forecasting
Weather Radar
Weather Satellites
Hazardous Weather: Hurricanes

The Hurricane: It's Nature, The Sentinal (Factory Insurance Association), p 3, Jul–Aug, 1971, in: Case Study #18, Hurricane Allen, South Texas Gulf Coast, August 10, 1980, Emergency Management Information Center, Learning Resource Center, National Emergency Training Center, Federal Emergency Management Agency, 16825 South Seton Ave, Emmitsburg, MD 21727, (800) 638-1821, (301) 447-1032.

Thygerson AL: Trauma Primer: Tornadoes, Emergency, 16:52, May 1984.

Tuffy B: 1001 Questions Answered About Storms and Other Natural Air Disasters, Dodd, Mead, New York, 1970.

Walker B: Earthquake, Time-Life Books, Alexandria, VA, 1982.

Weems JE: The Tornado, Doubleday & Co, 1977.

Yanev P: Peace of Mind in Earthquake Country, Chronicle Books, San Francisco, 1977.

DISASTER-RELEVANT ORGANIZATIONS

1. Volunteer Religious, Relief, and Welfare Organizations
2. University Programs and Disaster/Hazards Research
3. Other Sources of Disaster Information
4. Organizations that Offer Disaster Training

Volunteer Religious, Relief, and Welfare Organizations

AMERICAN NATIONAL RED CROSS
National Headquarters
18th and E Streets, NW
Washington, DC 20006

ANANDA MARGA UNIVERSAL RELIEF TEAM
854 Pearl Street
Denver, CO 80203

B'NAI B'RITH DISASTER RELIEF COMMISSION
1640 Rhode Island Avenue, NW
Washington, DC 20036

CHRISTIAN REFORMED WORLD RELIEF COMMITTEE
2850 Kalamazoo Avenue, SE
Grand Rapids, MI 49508

CHURCH OF THE BRETHREN DISASTER SERVICE
Brethren Service Center
Box 188
New Windsor, MD 21776

CHURCH WORLD SERVICE, DOMESTIC DISASTER COORDINATOR
475 Riverside Drive, Room 630
New York, NY 10027

EPISCOPAL CHURCH CENTER, PRESIDING BISHOPS FUND FOR W.R.
815 Second Avenue
New York, NY 10017

GOODWILL INDUSTRIES OF AMERICA
9200 Wisconsin Avenue, NW
Washington, DC 20014

LUTHERAN COUNCIL IN THE U.S.A., DOMESTIC DISASTER RESPONSE
360 Park Avenue South
New York, NY 10010

MENNONITE DISASTER SERVICE
21 South 12th Street
Akron, PA 17501

NATIONAL CATHOLIC DISASTER RELIEF COMMITTEE
Holy Redeemer Rectory
9705 Summit Avenue
Kensington, MD 20795

NATIONAL CONFERENCE OF CATHOLIC CHARITIES
1346 Connecticut Avenue, NW, Suite 307
Washington, DC 20036

SALVATION ARMY
National Headquarters
120 West 14th Street
New York, NY 10011

SEVENTH-DAY ADVENTISTS GENERAL CONFERENCE
6840 Eastern Avenue
Washington, DC 20012

SOUTHERN BAPTIST CONVENTION HOME MISSION BOARD
1350 Spring Street, NW
Atlanta, GA 30309

SOCIETY OF ST. VINCENT DE PAUL
Superior Council of the U.S.
4140 Lindell Boulevard
St. Louis, MO 63108

UNITED METHODIST CHURCH COMMITTEE ON RELIEF
475 Riverside Drive, Room 1470
New York, NY 10027

VOLUNTEERS OF AMERICA
340 West 85th Street
New York, NY 10024

University Programs and Disaster/Hazards Research

BATTELLE HUMAN AFFAIRS RESEARCH CENTERS
4000 NE 41st Street
Seattle, WA 98105

CENTER FOR ENVIRONMENTAL TOXICOLOGY
Michigan State University
East Lansing, MI

CENTER FOR PSYCHOLOGICAL RESPONSE IN DISASTER EMERGENCIES
New York Medical College
Valhalla, NY 10595

CENTER FOR TECHNOLOGY, ENVIRONMENT AND DEVELOPMENT
Clark University
Worcester, MA 01610

CENTER FOR URBAN AND REGIONAL STUDIES
Hickerson House 067A
University of North Carolina at Chapel Hill
Chapel Hill, NC 27514

COMMITTEE ON NATURAL DISASTERS
Academy of Engineering
National Academy of Sciences—National Research Council
Room JH-414
2101 Constitution Avenue
Washington, DC 20418

DISASTER MANAGEMENT PROGRAM
University of Wisconsin—Extension
Department of Engineering & Applied Science
432 North Lake Street
Madison, WI 53706

DISASTER PREPAREDNESS AND REHABILITATION PROJECT
East-West Center
University of Hawaii
1777 East-West Road
Honolulu, HI 96848

DISASTER RESEARCH CENTER
University of Delaware
Newark, DE 19716

DISASTER RESEARCH CENTER
Department of Civil Engineering
Texas Tech University
Lubbock, TX 79409

EARTHQUAKE ENGINEERING RESEARCH CENTER
University of California
1301 South 46th Street
Richmond, CA 94804

EARTHQUAKE ENGINEERING RESEARCH INSTITUTE
6431 Fairmount Avenue
El Cerrito, CA 94530

EMERGENCY ADMINISTRATION AND PLANNING DEGREE PROGRAM
North Texas State University
Denton, TX 76203

EMERGENCY COMMUNICATIONS RESEARCH UNIT
Carleton University
Colonel By Drive
Ottawa, Ontario
Canada K1S 5B6

HAZARDS ASSESSMENT LABORATORY
204 Aylesworth Hall
Colorado State University
Fort Collins, CO 80523

HAZARDS RESEARCH CENTER
Center for Public Affairs
Arizona State University
Tempe, AZ 85287

INDUSTRIAL CRISIS INSTITUTE
Management Department
School of Business
611 Tisch Hall
New York University
40 West 4th Street
New York, NY 10003

INSTITUTE FOR DISASTER PREPAREDNESS
MPA Program
311 South Spring Street
Los Angeles, CA 90013

INSTITUTE FOR RISK RESEARCH
University of Waterloo
Waterloo, Ontario
Canada N2L 3G1

INSTITUTE OF SAFETY AND SYSTEMS MANAGEMENT
ISSM 108
University of Southern California
Los Angeles, CA 90089-0021

INTERNATIONAL DISASTER INSTITUTE
1, Ferdinand Place
London, NW1 8EE, UK

MARYLAND INSTITUTE FOR EMERGENCY MEDICAL SERVICES
22 South Greene Street
Baltimore, MD 21201

NATIONAL DISASTER RESEARCH CENTER
Program of Policy Studies in Science and Technology
The George Washington University
Washington, DC 20052

NATURAL HAZARDS RESEARCH AND APPLICATIONS INFORMATION
CENTER
IBS #6, Campus Box 482
University of Colorado
Boulder, CO 80309-0482

RAND CORPORATION
1700 Main Street
Santa Monica, CA 90406-2138

RESEARCH COMMITTEE ON DISASTERS
Department of Sociology
Box 513, S-751 20 Uppsala
Sweden

RISK AND DECISION PROCESSES CENTER
The Wharton School
University of Pennsylvania
Philadelphia, PA 19104

RISK AND EMERGENCY MANAGEMENT PROGRAM
Center for Social and Urban Research
University of Pittsburgh
1617 Cathedral of Learning
Pittsburgh, PA 15260

WORLD ASSOCIATION FOR EMERGENCY AND DISASTER MEDICINE
Resuscitation Research Center
University of Pittsburgh
3434 Fifth Avenue
Pittsburgh, PA 15260

Other Sources of Disaster Information

ACADEMY OF HAZARD CONTROL MANAGEMENT
5010A Nicholson Lane
Rockville, MD 20852

ADVISORY COMMITTEE ON NONMILITARY RADIATION EMERGENCIES
American Medical Association
535 N. Dearborn Street
Chicago, IL 60610

ASSOCIATION OF AMERICAN WEATHER OBSERVERS
P.O. Box 455
Belvidere, IL 61008

ASSOCIATION OF BAY AREA GOVERNMENTS (ABAG)
P.O. Box 2050
Oakland, CA 94604-2050

AMERICAN CIVIL DEFENSE ASSOCIATION
P.O. Box 910
Starke, FL 32091

AMERICAN ASSOCIATION OF RAILROADS
Bureau of Explosives
1920 L Street, NW
Washington, DC 20036

AMERICAN PETROLEUM INSTITUTE
Hazardous Incident Training
2101 L Street, NW
Washington, DC 20037

AMERICAN PLANNING ASSOCIATION
1313 East 60th Street
Chicago, IL 60637-2891

AMERICAN PUBLIC WORKS ASSOCIATION, COUNCIL ON EMERGENCY MANAGEMENT
Manager of Emergency Management Services
1313 East 60th Street, Chicago, IL 60637

AMERICAN RADIO RELAY LEAGUE
225 Main Street
Newton, CT 06111

AMERICAN RESCUE DOG ASSOCIATION
Route 1, Box 161-K
Woodford, VA 22580

ASSOCIATION OF STATE DAM SAFETY OFFICIALS
P.O. Box 11910
Lexington, KY 40578

ASSOCIATION OF STATE FLOODPLAIN MANAGERS, INC.
P.O. Box 2051
Madision, WI 53701-2051

AUSTRALIAN NATIONAL DISASTERS ORGANIZATION
Australian Counter Disaster College
Mt. Macedon 3441
Victoria, Australia

AUSTRALIAN OVERSEAS DISASTER RESPONSE ORGANIZATION
P.O. Box K425
Haymarket, NSW 2000
Australia

BAY AREA REGIONAL EARTHQUAKE PREPAREDNESS PROJECT
101 8th Street, Suite 152
Oakland, CA 94607

BUILDING SEISMIC SAFETY COUNCIL
1015 15th Street, NW, Suite 700
Washington, DC 20005

CALIFORNIA EARTHQUAKE EDUCATION PROJECT
Lawrence Hall of Science
University of California
Berkeley, CA 94720

CALIFORNIA HAZARDOUS MATERIALS INVESTIGATORS ASSOCIATION
c/o Kern County District Attorney's Office
1215 Truxton Avenue
Bakersfield, CA 93301

CALIFORNIA SEISMIC SAFETY COMMISSION
1900 K Street
Sacramento, CA 95814

CALIFORNIA SPECIALIZED TRAINING INSTITUTE
P.O. Box 8104
San Luis Obispo, CA 93403

CENTER FOR MENTAL HEALTH—STUDIES OF EMERGENCIES
National Institute of Mental Health
U.S. Public Health Service
5600 Fishers Lane, Room 6C-12
Rockville, MD 20857

CENTRAL U.S. EARTHQUAKE CONSORTIUM
P.O. Box 367
Marion, IL 62959

CHEMICAL EDUCATION FOR PUBLIC UNDERSTANDING PROJECT
University of California
Lawrence Hall of Science
Berkeley, CA 94720

CHEMICAL MANUFACTURERS ASSOCIATION
2501 M Street, NW
Washington, DC 20037

CHLORINE INSTITUTE
70 West 40th Street
New York, NY 10018

CONGRESSIONAL TASK FORCE ON TOXIC EMERGENCIES
Environmental and Energy Study Conference
U.S. Congress
Washington, DC

DISASTER PREPAREDNESS OFFICE
National Weather Service
8060 13th Street
Silver Spring, MD 20910

DOCTORS FOR DISASTER PREPAREDNESS
P.O. Box 1057
Starke, FL 32091

EARTHQUAKE EDUCATION CENTER
Baptist College at Charleston
P.O. Box 10089
Charleston, SC 29411

EARTHQUAKE EDUCATION PLANNING PROJECT
Federal Emergency Management Agency
50 C Street, SW
Washington, DC 20472

EARTHQUAKE ENGINEERING RESEARCH CENTER
University of California
1301 South 46th Street
Richmond, CA 94804

EARTHQUAKE INFORMATION SERVICE
U.S. Geological Survey
MS 967, Box 25046
Denver Federal Center
Denver, CO 80225

EMERGENCY MANAGEMENT INSTITUTE
P.O. Box 102
Sterling, VA 22170

EMERGENCY MANAGEMENT INFORMATION CENTER
Learning Resources Center
National Emergency Training Center
Federal Emergency Management Agency
16825 South Seton Avenue
Emmitsburg, MD 21727

EMERGENCY MANAGEMENT INFORMATION SERVICES
25 McLean Place
Indianapolis, IN 46202-1322

EMERGENCY PREPAREDNESS—CANADA
141 Laurier Avenue, West, 2nd Floor
Ottawa, Canada
K1A 0W6

EMERGENCY RESPONSE INSTITUTE
319 Olive Street
Cashmere, WA 98815

HAZARD MANAGEMENT GROUP
Oak Ridge National Laboratory
Energy Division
P.O. Box X
Oak Ridge, TN 37831

HAZARDOUS MATERIALS ADVISORY COUNCIL
1100 17th Street, NW, Suite 908
Washington, DC 20036

HAZARDOUS MATERIALS INFORMATION EXCHANGE
Federal Emergency Management Agency
Technological Hazards Division
500 C Street, SW
Washington, DC 20472

INSTITUTE OF MAKERS OF EXPLOSIVES
1575 Eye Street, NW, Suite 550
Washington, DC 20005

INTERNATIONAL ASSOCIATION OF FIRE CHIEFS
1329 18th Street, NW
Washington, DC 20036

INTERNATIONAL CITY MANAGERS ASSOCIATION
1120 G Street, NW
Washington, DC 20005

INTERNATIONAL SOCIETY ON DISASTER MEDICINE
P.O. Box CH-1213 Petit-Lancy 2
Switzerland

INTERNATIONAL TSUNAMI (Tidal Wave) INFORMATION CENTER
P.O. Box 50027
Honolulu, HI 96850

NATIONAL AGRICULTURAL CHEMICALS ASSOCIATION
1155 15th Street, NW, Suite 514
Washington, DC 20005

NATIONAL ASSOCIATION FOR SEARCH AND RESCUE
P.O. Box 50178
Washington, DC 20004

NATIONAL ASSOCIATION OF URBAN FLOOD MANAGEMENT
AGENCIES
1015 18th Street, NW, Suite 1002
Washington, DC 20036

NATIONAL COORDINATING COUNCIL ON EMERGENCY
MANAGEMENT
Public Affairs Office
National Headquarters
3125 Beltline Boulevard, Suite 101
Columbia, SC 29204

NATIONAL DISASTER MEDICAL SYSTEM
Parklawn Building, Room 16A-54
5600 Fishers Lane
Rockville, MD 20857

NATIONAL EMERGENCY MANAGERS ASSOCIATION
c/o State Emergency Management Office
Division of Military and Naval Affairs
Public Security Building, State Campus
Albany, NY 12226

NATIONAL EMERGENCY TRAINING CENTER
Federal Emergency Management Agency
P.O. Box 225
Emmitsburg, MD 21727

NATIONAL EMS CLEARINGHOUSE
c/o The Council of State Governments
P.O. Box 11910
Iron Works Pike
Lexington, KY 40578

NATIONAL FIRE PROTECTION ASSOCIATION
Batterymarch Park
Quincy, MA 02269

NATIONAL HURRICANE CENTER/NOAA
Gable 1 Tower, Room 631
1320 South Dixie Highway
Coral Gables, FL 33146

NATIONAL INFORMATION SERVICE FOR EARTHQUAKE ENGINEERING
379 Davis Hall
University of California
Berkeley, CA 94720

NATIONAL INSTITUTE OF MENTAL HEALTH
Center for Mental Health Studies of Emergencies
Parklawn Building, Room 7C-02
5600 Fishers Lane
Rockville, MD 20857

NATIONAL TOXICOLOGY PROGRAM
Department of Health and Human Services
P.O. Box 12233
Research Triangle Park, NC 27709

NATIONAL TRANSPORTATION SAFETY BOARD
800 Independence Avenue, SW
Washington, DC 20594

NATURAL AND MAN-MADE HAZARD MITIGATION
Directorate of Engineering
National Science Foundation
1800 G Street, NW
Washington, DC 20550

NATURAL DISASTER RECOVERY AND MITIGATION RESOURCE
REFERRAL SERVICE
Academy for Contemporary Problems
400 North Capitol Street, NW, Suite 390
Washington, DC 20001

NATURAL DISASTER RESOURCE REFERRAL SERVICE
P.O. Box 2208
Arlington, VA 22202

OFFICE OF THE UNITED NATIONS DISASTER RELIEF COORDINATOR
Palais des Nations, CH-1211
Geneva 10
Switzerland

NORTHWEST EMERGENCY PREPAREDNESS ASSOCIATION
Box 70097
Bellevue, WA 98007

NUCLEAR REGULATORY COMMISSION
Incident Response Branch
Office of Inspection and Enforcement
Maryland National Bank Building
Washington, DC 20555

PAN AMERICAN CARIBBEAN DISASTER PREPAREDNESS
AND PREVENTION PROJECT
Factory Road
Antigua, West Indies

PAN AMERICAN HEALTH ORGANIZATION
Emergency Preparedness and Disaster Relief Program
525 23rd Street, NW
Washington, DC 20037

PARTNERS OF THE AMERICAS
Emergency Preparedness Program
1424 K Street, NW, Suite 700
Washington, DC 20005

RADIO EMERGENCY ASSOCIATED CITIZENS TEAMS (REACT)
75 East Wacker Drive
Chicago, IL 60601

RESEARCH ALTERNATIVES
10221 Arizona Circle
Bethesda, MD 20034

SCHOOL EARTHQUAKE SAFETY AND EDUCATION PROJECT
State Seismologist
Geophysics Department, AD-50
University of Washington
Seattle, WA 98195

SOCIETY FOR RISK ANALYSIS
1340 Old Chain Bridge Road, Suite 300
McLean VA 22101

SOUTHERN CALIFORNIA EARTHQUAKE PREPAREDNESS PROJECT
6858 Van Nuys Boulevard
Van Nuys, CA 91405

SPILL CONTROL ASSOCIATION OF AMERICA
1515 North Park Plaza
Southfield, MI 48075

TENNESSEE EARTHQUAKE INFORMATION CENTER
Memphis State University
Memphis, TN 38152

THE TSUNAMI SOCIETY
P.O. Box 8523
Honolulu, HI 96815

U.S. COAST GUARD
National Response Center
2100 Second Street, SW
Washington, DC 20593

U.S. ENVIRONMENTAL PROTECTION AGENCY
Emergency Response Division
WH 548 B, 401 M Street, SW
Washington, DC 20460

U.S. ARMY CORPS OF ENGINEERS
Water Resources Support Center
Casey Building, No. 2594
Ft. Belvoir, VA 22060

U.S. DEPARTMENT OF ENERGY
Public Information Office
1E 218 Forrestal Building
1000 Independence Avenue, SW
Washington, DC 20585

U.S. DEPARTMENT OF TRANSPORTATION
Materials Transportation Bureau
Office of Hazardous Materials Regulations
400 Seventh Street, SW
Washington, DC 20472

U.S. DEPARTMENT OF TRANSPORTATION
National Standards Division
Bureau of Motor Carrier Safety
400 Seventh Street, SW
Washington, DC 20472

Organizations that Offer Disaster Training

ALM ENTERPRISES (Hazmat)
P.O. Box 20912
El Cajon, CA 92021

AMERICAN TRUCKING ASSOCIATION (Hazmat)
The Operations Council
1616 P Street, NW
Washington, DC 20036

ASSOCIATION OF BAY AREA GOVERNMENTS
101 8th Street
Oakland, CA 94607

CALIFORNIA STATE FIRE MARSHAL
Supervisor, Fire Service Training and Education Program
7171 Bowling Drive, Suite 600
Sacramento, CA 95023

CALIFORNIA SPECIALIZED TRAINING INSTITUTE
P.O. Box 8104
San Luis Obispo, CA 93403-8104

CANADIAN CHEMICAL PRODUCER'S ASSOCIATION
Suite 805, 350 Sparks Street
Ottawa, Ontario
Canada K1R 7S8

CELANESE CORPORATION (Hazmat)
Fire Training Center
Dean of Extension Service
York Technical College
Rock Hill, SC 29730

CHEMICAL MANUFACTURERS ASSOCIATION
2501 M Street, NW
Washington, DC 20037

COLORADO TRAINING INSTITUTE (Hazmat)
1001 East 62nd Avenue
Denver, CO 80216

DAVID FRANK ASSOCIATES (Hazmat)
416 South Rolling Road
Catonsville, MD 21228

EMERGENCY ACTION, INC. (Hazmat)
P.O. Box 10661
Charleston, SC 29411

EMERGENCY RESPONSE INSTITUTE
319 Olive Street
Cashmere, WA 98815

ENVIRONMENT CANADA
Technology Development and Tech. Services Branch
Ottawa, Ontario, Canada K1A 1C8

ENVIRONMENTAL HAZARDS MANAGEMENT INSTITUTE (Hazmat)
P.O. Box 283
Portsmouth, NH 03801

ENVIRONMENTAL SAFETY AND DESIGN (Hazmat)
P.O. Box 34207
Memphis, TN 38134

FEDERAL EMERGENCY MANAGEMENT AGENCY
National Training Center
16825 South Seton Avenue
Emmitsburg, MD 21727-8995

FEDERAL EMERGENCY MANAGEMENT AGENCY
National Fire Academy
16825 South Seton Avenue
Emmitsburg, MD 21727-8995

FIRE RESCUE CONSULTANTS (Hazmat)
P.O. Box 5703
Rockville, MD 20855

GOVERNMENT SERVICES INSTITUTE
P.O. Box 866
Point Lookout, MO 65726

HAZARDOUS MATERIALS CONTROL RESEARCH INSTITUTE
9300 Columbia Boulevard
Silver Spring, MD 20910

HAZARDOUS RISK ADVISORY COMMITTEE OF NASHVILLE (Hazmat)
Metro Civil Defense
Floor 7-M, Metro Courthouse
Nashville, TN 37201

HAZTECH INTERNATIONAL
c/o Conference Coordinator
6143 South Willow Drive, Suite 100
Englewood, CO 80111-5115

INTERNATIONAL CONGRESS OF ENVIRONMENTAL PROFESSIONALS
P.O. Box 178
Techny, IL 60082

INTERNATIONAL SOCIETY OF FIRE SERVICE INSTRUCTORS
20 Main Street
Ashland, MA 01721

J.T. BAKER CHEMICAL CO. (Hazmat)
Office of Safety Training
222 Red School Lane
Phillipsburg, NJ 08865

MASSACHUSETTS FIREFIGHTING ACADEMY (Hazmat)
Coordinator, LNG/LPG Firefighting School
59 Horse Pond Road
Sudbury, MA 01776

NATIONAL ASSOCIATION FOR SEARCH AND RESCUE
P.O. Box 50178
Washington, DC 20004

NATIONAL FIRE PROTECTION ASSOCIATION
Division of Continuing Education
Batterymarch Park
Quincy, MA 02269

NATIONAL MINE HEALTH AND SAFETY ACADEMY
P.O. Box 1166
Beckley, WV 25801

NATIONAL SPILL CONTROL SCHOOL (Hazmat)
Corpus Christi State University
6300 Ocean Drive
Corpus Christi, Texas 78412

OSHA TRAINING INSTITUTE
1555 Times Drive
Des Plaines, IL 60018

ROBERTS ENVIRONMENTAL SERVICES, INC.
P.O. Box 10093
Eugene, OR 97440

SOCIETY FOR COMPUTER SIMULATION
P.O. Box 17900
San Diego, CA 92117-7900

SAFETY SYSTEMS, INC. (Hazmat)
P.O. Box 8463
Jacksonville, FL 32239

TEXAS A & M UNIVERSITY (Hazmat)
Oil and Hazardous Material Control Training Division
Texas Engineering Extension Service
The Texas A & M University System
F.E., Drawer K
College Station, TX 77843

TRANSPORTATION SAFETY INSTITUTE (Department of Transportation
Hazmat courses; restricted to state and federal employees)
Mike Monroney Aeronautical Center
6500 South MacArthur Boulevard
Oklahoma, OK 73125

UNIVERSITY OF WISCONSIN
Department of Engineering
Professional Development
Madison, WI 53706

U.S. COAST GUARD
Pollution Response Branch
U.S.C.G. Headquarters (G-WER-2)
2100 2nd Street, SW
Washington, DC 20593

U.S. DEPARTMENT OF TRANSPORTATION
Transportation Safety Institute
Hazardous Materials Safety Program
DMA-604
6500 South MacArthur Boulevard
Oklahoma City, OK 73125

U.S. ENVIRONMENTAL PROTECTION AGENCY
Hazardous Materials Incident Response Training Program
26 West St. Clair Street
Cincinnati, OH 45268

G

DISASTER-RELEVANT PUBLICATIONS

1. Disaster-Relevant Periodicals

2. Catalogues, Publication Lists, Computer Data Bases

3. Disaster-Related Bibliographies

Disaster-Relevant Periodicals

ALBERTA DISASTER SERVICES NEWS AND NOTES
Alberta Disaster Services Agency
10320 146th Street, Edmonton
Alberta, Canada T5N 3A2
FREE

AMERICAN HEAT (fire service video periodical)
8001 Clayton Road
St. Louis, MO 63117

AMERICAN WEATHER OBSERVER
Association of American Weather Observers
P.O. Box 455
Belvidere, IL 61008
Monthly
subscription part of $16/yr. membership

ANNALS OF EMERGENCY MEDICINE
American College of Emergency Medicine
P.O. Box 619911
Dallas, TX 75261-9911

AODRO NEWSLETTER
Australian Overseas Disaster Response Organization
P.O. Box K425
Haymarket, NSW 2000
Australia
Bimonthly
subscription part of membership fee

APCO BULLETIN
Communications Officers, Inc. (APCO)
P.O. Box 669
New Smyrna Beach, FL 32069

ASDSO NEWSLETTER
Association of State Dam Safety Officials
P.O. Box 11910
Lexington, KY 40578
Quarterly
subscription part of $10/yr membership fee

ASFPM NEWS & VIEWS
Association of State Floodplain Managers, Inc.
P.O. Box 2051
Madison, WI 53701-2051
Bimonthly
subscription part of membership fee

BUILDING SEISMIC SAFETY COUNCIL BULLETIN
Building Seismic Safety Council
1015 15th Street, NW, Suite 700
Washington, DC 20005
Bimonthly
subscription part of annual membership

CALEEP NEWS
California Earthquake Education Project
Lawrence Hall of Science
University of California
Berkeley, CA 94720
Three times a year
FREE

CARIBBEAN DISASTER PREPAREDNESS NEWSLETTER
Pan Caribbean Disaster Preparedness and Prevention Project
P.O. Box 1399
St. John's Antigua, West Indies
Quarterly

CICS NEWSLETTER
Colorado Incident Command System
P.O. Box 271
Manitou Springs, CO 80829

CIVIL AIR PATROL NEWS
U.S. Air Force
Civil Air Patrol Headquarters
Maxwell AFB, AL 36112
Bimonthly
$2.00

DANGEROUS GOODS NEWSLETTER
Transport Canada
Tower B, Place De Ville
Ottawa, Ontario
Canada K1A ON5
FREE

DISASTER LIFELINES
Northwest Emergency Preparedness Association
P.O. Box 70097
Bellevue, WA 98007

DISASTER MANAGEMENT
Joint Assistance Centre
H-65, South Extension Part I
New Delhi 110049, India
Quarterly
$20 a year

DISASTER PREPAREDNESS IN THE AMERICAS
Pan American Health Organization
525 23rd Street, NW
Washington, DC 20037
Monthly
FREE

DISASTER PREPAREDNESS REPORT
National Weather Service
8060 13th Street, Room 1326
ATTN: W/Om11x1—Lorraine Brown
Silver Spring, MD 29010
Quarterly
FREE

DISASTER PREPAREDNESS SERIES
Pan American Health Organization
525 23rd Street, NW
Washington, DC 20037

DISASTERS, THE INTERNATIONAL JOURNAL OF DISASTER STUDIES
AND PRACTICE
International Disaster Institute
1, Ferdinand Place
London NW1 8EE, U.K.
Quarterly
$50/vol. indiv.; $75/vol. instit.

EARTHQUAKE INFORMATION BULLETIN
Superintendent of Documents
U.S. Government Printing Office
Washington, DC 20402
Bimonthly
$15/yr. domestic; $18.75/yr. foreign

EARTHQUAKE PREDICTION RESEARCH
D. Reidel Publishing Company
190 Old Derby Street
Hingham, MA 02043
Quarterly
$34/yr. indiv.; $72/yr. instit.

EARTHQUAKE SPECTRA
Earthquake Engineering Research Institute
6431 Fairmount Avenue
El Cerrito, CA 94530
Quarterly
$50/yr. indiv.; $90/yr. instit.; +$15 for foreign

EERC NEWS
Earthquake Engineering Research Center
University of California
1301 South 46th Street
Richmond, CA 94804
Quarterly
FREE

EERI NEWSLETTER
Earthquake Engineering Research Institute
6431 Fairmount Avenue
El Cerrito, CA 94530
Monthly
Subscription part of annual membership

EMERGENCY MANAGEMENT DISPATCH
Department of Political Science
University of Delaware
Newark, Del 19716

EMERGENCY MANAGEMENT TODAY
25 McLean Place
Indianapolis, IN 46202

EMERGENCY MEDICAL SERVICES: THE JOURNAL OF EMERGENCY CARE
AND TRANSPORTATION
7628 Densmore Ave
Van Nuys, CA 91406-2088

EMERGENCY PREPAREDNESS DIGEST
Emergency Preparedness—Canada (Canada's equivalent of FEMA)
141 Laurier Avenue, West, 2nd Floor
Ottawa, Ontario
Canada
K1A 0W6
Quarterly
FREE

EMERGENCY PREPAREDNESS NEWS
Business Publishers, Inc.
951 Pershing Drive
Silver Spring, MD 20910

FEMA NEWSLETTER
Federal Emergency Management Agency
Washington, DC 20472
Bimonthly
FREE

FIRE ENGINEERING
P.O. Box 1260
Tulsa, OK 74101

FIRE CHIEF: ADMINISTRATION/TRAINING/OPERATIONS
6255 Barfield Road
Atlanta, GA 30328

FIRE COMMAND
National Fire Protection Association
Batterymarch Park
Quincy, MA 02269

FLOOD REPORT
EMMA, Inc.
P.O. Box 11259
Alexandria, VA 22312
Monthly
$89 a year

GROUND FAILURE
National Research Council
Committee on Ground Failure Hazards
2101 Constitution Avenue
Washington, DC 20418
Three times a year

HAZARD MONTHLY
P.O. Box 8438
Rockville, MD 20856
Monthly
$26 a year; more for institutions and foreign subscribers

HAZARDOUS MATERIALS INTELLIGENCE REPORT
World Information Systems
P.O. Box 535, Harvard Square Station
Cambridge, MA 02238

HAZARDOUS MATERIALS NEWSLETTER
U.S. Department of Transportation
Materials Transportation Bureau
Research and Special Program Administration
Washington, DC 20590
FREE

HAZARDOUS MATERIALS NEWSLETTER
P.O. Box 204
Barre, VT 05641
Bimonthly
$35 a year domestic

HAZARDOUS MATERIALS TRANSPORTATION
Cahners Publishing Company
221 Columbus Avenue
Boston, MA 02116
$195 a year

HAZARDOUS WASTE NEWS
Business Publishers, Inc.
951 Pershing Drive
Silver Spring, MD 20910
Weekly
$257/yr. and $10 postage

HAZARDOUS WASTE REPORT
Aspen Systems, Inc.
1600 Research Boulevard
Rockville, MD 20850
Biweekly
$260 a year

INDUSTRIAL CRISIS QUARTERLY
Industrial Crisis Institute, Inc.
100 Bleecker Street, Suite 2B
New York, NY 10012

INTERNATIONAL CIVIL DEFENSE
International Civil Defense Organization
10-12 chemin de Surville
Ch-1213
Petit-Lancy
Geneva, Switzerland
Monthly
38 Swiss Francs a year

INTERNATIONAL JOURNAL OF MASS EMERGENCIES AND DISASTERS
Engelbrekt Distribution
Engelbreksgatan 13
S-114 32 Stockholm, Sweden
Three times a year
subscription part of $20/yr membership fee in the Research Committee on
Disasters

INTERNATIONAL REVIEW OF THE RED CROSS
International Committee of the Red Cross
17 Avenue de la Paix
1211 Geneva 1, Switzerland
Monthly
$12.00

JOURNAL OF CIVIL DEFENSE
American Civil Defense Association
P.O. Box 910
Starke, FL 32091

JOURNAL OF EMERGENCY MEDICAL SERVICES
P.O. Box 27966
San Diego, CA 92128
Monthly
$15.95 a year (domestic)

JOURNAL OF THE WORLD ASSOCIATION FOR EMERGENCY AND
DISASTER MEDICINE
World Association for Emergency and Disaster Medicine
Resuscitation Research Center
University of Pittsburgh
3434 Fifth Avenue
Pittsburgh, PA 15260

MACEDON DIGEST
Australian Natural Disasters Organization
Counter Disaster College
Mt. Macedon 3441
Victoria, Australia

MISSOURI DISASTER PLANNING AND OPERATIONS NEWSLETTER
Disaster Planning and Operations Office
1717 Industrial Drive
Jefferson City, MO 65101
Bimonthly
FREE

NATIONAL EMERGENCY TRAINING GUIDE
1819 Mark Street, NE
Olympia, WA 98506
Biannually
$37.50 a year

NATURAL HAZARD OBSERVER
Natural Hazards Research and Applications Information Center
I.B.S. #6, Campus Box 482
University of Colorado
Boulder, CO 80309

NAUFMA MONTHLY NEWS
National Association of Urban Flood Management Agencies
1015 18th Street, NW, Suite 1002
Washington, DC 20036
Monthly
Subscription part of annual dues

NETMA NEWS
National Emergency Training Center
Federal Emergency Management Agency
P.O. Box 225
Emmitsburg, MD 21727

NETWORK NEWSLETTER
Environmental and Societal Impacts Group
National Center for Atmospheric Research
P.O. Box 3000
Boulder, CO 80307
Quarterly
FREE

NETWORKS: EARTHQUAKE PREPAREDNESS NEWS
Bay Area Regional Earthquake Preparedness Project
Metro-Center
101 8th Street, Suite 152
Oakland, CA 94607

NUCLEAR WASTE NEWS
Business Publishers, Inc.
951 Pershing Drive
Silver Spring, MD 20910
Biweekly
$237 a year plus $5.20 postage

RESPONSE
National Association for Search and Rescue
P.O. Box 27966
San Diego, CA 92128
Bimonthly
$14.95 a year domestic

RISK ABSTRACTS
Institute for Risk Research
University of Waterloo
Waterloo, Ontario
Canada N2L 3G1
Quarterly
$65 a year U.S.; $80 a year Canada

RISK ANALYSIS
Society for Risk Analysis
1340 Old Chain Bridge Road, Suite 300
McLean, VA 22101
Quarterly
Subscription included in $30 a year membership fee

SAR (Search and Rescue) DOG ALERT
P.O. Box 39
Somerset, CA 95684

SCIENCE OF TSUNAMI HAZARDS
The Tsunami Society
Box 8523
Honolulu, HI 96815
Biannually
Subscription part of $25 a year membership fee

SIREN
Department of Law and Public Safety
Division of Civil Defense—Disaster Control
Eggert Crossing Road
Box 979
Trenton, NJ 08625
Two issues a year
FREE

SPILL TECHNOLOGY NEWSLETTER
Environmental Protection Service
Department of the Environment
Ottawa, Ontario
Canada K1A 1C8
FREE

STORM DATA
National Climatic Data Center
National Environmental Satellite, Data, and Information Service
NOAA
Federal Building
Asheville, NC 28801-2696
Monthly
$15 a year

TOXIC MATERIALS NEWS
Business Publishers, Inc.
951 Pershing Drive
Silver Spring, MD 20910
Weekly
$257 plus $10.00 postage

TOXIC MATERIALS TRANSPORT
Business Publishers, Inc.
951 Pershing Dr.
Silver Spring, MD 20910
Biweekly
$157 plus postage

TRIAGE
Doctors for Disaster Preparedness
P.O. Box 1057
Starke, FL 32091

TSUNAMI NEWSLETTER
International Tsunami Information Center
P.O. Box 50027
Honolulu, HI 96850
Biannually

UNDRO NEWS
Office of the United Nations Disaster Relief Coordinator (UNDRO)
Palais des Nations, CH-1211
Geneva 10
Switzerland

UNSCHEDULED EVENTS
Research Committee on Disasters
c/o Joanne Nigg
Office of Hazard Studies
Arizona State University
Tempe, AZ 85287
Three times a year
Subscription part of $20 annual membership fee

UPDATE
Southern California Earthquake Preparedness Project
600 South Commonwealth Avenue, Suite 1100
Los Angeles, CA 90005
Quarterly

U.S. CIVIL DEFENSE COUNCIL BULLETIN
U.S. Civil Defense Council
P.O. Box 370
Portsmouth, VA 23705
Monthly
$4.00

VOLCANO NEWS
320 East Shore Drive
Kemah, TX 77565
Quarterly
$7 a year domestic; $11 a year foreign

Catalogues, Publication Lists, Computer Data Bases

CHEMTREC
24-hour Hazardous Chemical Information Hotline
Chemical Manufacturers Association
2501 M Street, NW
Washington, DC 20037
(800) 424-9300
(202) 483-7616 (Hawaii, Alaska, and DC)

CIS, INC.
Fein Marquart Associates
Chemical Databases
7215 York Road
Baltimore, MD 21212
(800) 247-8737

CIVIL EMERGENCY PREPAREDNESS EDUCATION AND TRAINING
PROGRAMS
National AudioVisual Center
Information Services JU
Washington, DC 20409

DISASTER RESEARCH CENTER
Publication List
University of Delaware
Newark, DE 19716

EMERGENCY MANAGEMENT INFORMATION CENTER
Learning Resource Center Library
Disaster Case Studies
Federal Emergency Management Agency
National Emergency Training Center
16825 South Seton Avenue
Emmitsburg, MD 21727
(800) 638-1821

EMERGENCY RESPONSE INSTITUTE
National Emergency Training Guide
1819 Mark Street, NE
Olympia, WA 98506

FEDERAL EMERGENCY MANAGEMENT AGENCY
Publications Catalogue
P.O. Box 8181
Washington, DC 20024

FEDERAL EMERGENCY MANAGEMENT AGENCY
HAZMAT Information Exchange (computer data base)
500 C Street, SW
Washington, DC 20472
(800)-752-6367, (800)-367-9592 (Illinois residents)
(312) 972-3275 or FTS 972-3275 for computer modem access

INFORMATION CONSULTANTS, INC
Chemical Information System Databases
1133 15th Street, NW
Washington, DC 20005

INTERNATIONAL FIRE SERVICE TRAINING ASSOCIATION
Fire Service Training Materials Catalog
Fire Protection Publications
Oklahoma State University
Stillwater, OK 74078-9987
(800) 654-4055

LAB SAFETY SUPPLY
Safety Equipment Catalogue (includes hazmat equipment and books)
P.O. Box 1368
Janesville, WI 53547-1368
(800) 356-0783

NATIONAL FIRE PROTECTION ASSOCIATION
The NFPA Catalog
Batterymarch Park
Quincy, MA 02269
(800) 344-3555

NATIONAL INSTITUTE OF MENTAL HEALTH
Center for Mental Health Studies of Emergencies
Publications List
Parklawn Building, Room 7C-02
Rockville, MD 20857

NATIONAL INTERAGENCY INCIDENT MANAGEMENT SYSTEM (NIIMS)
Publications Management System Catalog
(Publications related to NIIMS and the Incident Command System)
National Wildfire Coordinating Group
Boise Interagency Fire Center
3905 Vista Avenue
Boise, ID 83705

NATIONAL LIBRARY OF MEDICINE
Toxicological Data Network (TOXNET)
Specialized Information Services
Biomedical Files Implementation Branch
8600 Rockville Pike
Bethesda, MD 20894
(301) 496-6531
(301) 496-1131

NATURAL HAZARDS RESEARCH AND APPLICATIONS INFORMATION
CENTER
Publications Lists
Annual Bibliographies on Disaster Articles
Institute of Behavioral Science #6
Campus Box 482
University of Colorado
Boulder, CO 80309

RESEARCH ALTERNATIVES
966 Hungerford Drive, Suite 31B
Rockville, MD 20850

SEISMIC SAFETY COMMISSION
Publications List
1900 K Street
Sacramento, CA 95814-4186
(916) 322-4917

SPECIALIZED PUBLICATION SERVICES, INC.
Fire Service Directory of Training and Information Sources
151 First Avenue, Suite 102
New York, NY 10003-9808

U.S. COAST GUARD PUBLICATIONS
Subject Bibliography, SB-263
Superintendent of Documents
U.S. Government Printing Office
Washington, DC 20402

U.S. DEPARTMENT OF DEFENSE
Hazardous Materials Information System
Defense General Supply Center
Richmond, VA 23297
(804) 275-3104

Disaster-Related Bibliographies

Ahearn FL, and Cohen RE: Disasters and mental health: an annotated bibliography, 1983, Available from: National Institute of Mental Health, 5600 Fishers Lane, Rockville, MD.

Bibliography of dissertations and theses on disaster phenomena, Unscheduled Events 2, Winter, 1968.

Bay Area Regional Earthquake Preparedness Project, Earthquake preparedness and public information: an annotated bibliography, 1985. Available from: BAREPP 85-2, MetroCenter, 101 Eighth St, Suite 152, Oakland, CA 94607.

Disaster Research Group, Field studies of disaster behavior: an inventory, Publication 886, National Academy of Sciences—National Research Council, Washington, DC, 1961.

Emergency Response Institute, National emergency training and information guide, 1988. Available from: 1819 Mark St, NE, Olympia, WA 98506.

Fitzsimmons AR: Natural hazards and land use planning: an annotated bibliography, 1984. Available from: Natural Hazards Research and Applications Information Center, IBS #6, Campus Box 482, University of Colorado, Boulder, CO 80309, $5.50.

Fritz CE, et al: An inventory of field studies on human behavior in disasters, National Academy of Sciences—National Research Council, Washington, DC, Aug 15, 1959.

Golant S: Human behavior before the disaster: a selected annotated bibliography, 1969. Available from: Natural Hazards Research and Applications Information Center, IBS #6, Campus Box 482, University of Colorado, Boulder, CO 80309.

Harnly CD, and Tyckson DA: Mt. St. Helens, the 1980 eruptions: a bibliography. Available from: Vance Bibliographies, Public Administra-tion Series, Bibliography #P 786, Miami University, Miami, FL.

Mitchell JK: A selected bibliography of coastal erosion, protection and related human activity in North America and the British Isles, 1968. Available from: Natural Hazards Research and Applications Information Center, IBS #6, Campus Box 482, University of Colorado, Boulder, CO 80309.

Morton DR: Bibliography on natural disaster recovery and reconstruction, 1979. Available from: Natural Hazards Research and Applications Information Center, IBS #6, Campus Box 482, University of Colorado, Boulder, CO 80309, $2.00.

Morton DR: A selected bibliography on disaster planning and simulation, 1981. Available from: Natural Hazards Research and Applications Information Center, IBS #6, Campus Box 482, University of Colorado, Boulder, CO 80309, $2.00.

Morton DR: A selected, partially annotated bibliography of recent (1982–1983) natural hazards publications, 1984. Available from: Natural Hazards Research and Applications Information Center, IBS #6, Campus Box 482, University of Colorado, Boulder, CO 80309.

Morton DR: A selected, partially annotated bibliography of recent (1984–1985) natural hazards publications, 1986. Available from: Natural Hazards Research and Applications Information Center, IBS #6, Campus Box 482, University of Colorado, Boulder, CO 80309.

Natural Hazards Research and Applications Information Center, Annual bibliographies of articles, reports, and studies pertaining to societal aspects of natural hazards and disasters (annually). Available from: IBS #6, Campus Box 482, University of Colorado, Boulder, CO 80309.

National Technical Information Service: Disasters: effects, preparedness, assessment, and recovery, 1976—February, 1982. Citations from the NTIS Data Base, Springfield, VA.

Pope T, and Wenger D: TMI [Three Mile Island] in the literature: a partially annotated bibliography, Int J Mass Emerg and Disas 2(1):197, Nov 1984. Available from: International Library, PO Box 1839, S-751 01, Uppsala, Sweden.

Quarantelli EL: A selected annotated bibliography of social science studies on disasters, Amer Behav Sci 13(3):452–456, Jan/Feb 1970.

Quarantelli EL: A 100 item annotated bibliography on disaster and disaster planning, 1980. Available from: Disaster Research Center, University of Delaware, Newark, 19716.

Quarantelli EL: Inventory of disaster field studies in the social and behavioral sciences: 1919–1979, Miscellaneous Report No. 32, 1982. Available from: Disaster Research Center, University of Delaware, Newark, 19716.

Rayner J: Annotated bibliography on disaster research, Human Organization 16(2):30, Summer, 1957.

Relph EC, and Goodwillie SB: Annotated bibliography on snow and ice problems, 1968. Available from: Natural Hazards Research and Applications Information Center, IBS #6, Campus Box 482, University of Colorado, Boulder, CO 80309.

Reynolds S, and Wright JE: A selective literature review of disaster medical services, working paper #64, 1976. Available from: Disaster Research Center, University of Delaware, Newark, DE 19716.

Specialized Publication Services, Fire service directory of training and information sources, 1987. Available from: 151 First Avenue, Suite 102, New York, NY 10003-9808.

Wilson E: A selected annotated bibliography and guide to sources of information and planning for and responses to chemical emergencies, J Haz Mat 4(4):373, 1981.

Young ME: Disasters (A bibliography with abstracts), 1964–August 1975, National Technical Information Service, Springfield, VA, 1975.

Young ME: Disasters: effects and countermeasures, 1964–October 1976, National Technical Information Service, Springfield, VA, 1976.

Young ME: Disasters: effects and countermeasures, vol 2, 1976–1977, National Technical Information Service, Springfield, VA, 1977.

Young ME: Disasters: effects and countermeasures, vol 2, 1976–October 1978, National Technical Information Service, Springfield, VA, 1978.

Young ME: Disasters: effects, preparedness, assessment and recovery, vol 1, 1964–1975 (A bibliography with abstracts), National Technical Information Service, Springfield, VA, 1979.

Young ME: Disasters: effects, preparedness, assessment and recovery, 1976–December, 1980, Citations from the NTIS Data Base, National Technical Information Service, Springfield, VA, 1981.

NOTE: The author has attempted to provide accurate and up-to-date information for this listing. Any inconvenience caused by changes in addresses or prices is regrettable, but sometimes unavoidable.

ADDITIONAL READING ON THE GENERAL TOPIC OF DISASTER MANAGEMENT

Adams CA: Law: its effects on interorganizational authority structures in post-disaster responses, Tech Rep no 9, SAR Research Project, 1981. Available from: Department of Sociology, University of Denver, University Park, Denver, CO 80208.

Adams CR: Search and rescue efforts following the Wichita Falls tornado, Tech Rep no 4, SAR Research Project, 1981. Available from: Department of Sociology, University of Denver, University Park, Denver, CO 80208.

Adams CR, and Drabek TE: Legal issues in natural disaster responses, A paper presented at the Annual Meeting of the Pacific Sociological Association, San Francisco, April, 1980. Available from: Department of Sociology, University of Denver, University Park, Denver, CO 80208.

Adams DS: Emergency actions and disaster reactions: an analysis of the Anchorage Public Works Department in the 1964 Alaskan earthquake, Disaster Research Center Monograph Series no 5, 1969. Available from: Disaster Research Center, University of Delaware, Newark, 19716.

American Medical Association: A guide to the hospital management of injuries arising from exposure to or involving ionizing radiation, 1984. Available from: American Medical Association, Department—OP 335, PO Box 10946, Chicago, IL 60610.

Bay Area Regional Earthquake Preparedness Project: City comprehensive earthquake pre-paredness planning guidelines, 1985. Available from: BAREPP, MetroCenter, 101 Eighth St, Suite 152, Oakland, CA 94607.

Bay Area Regional Earthquake Preparedness Project: Corporate comprehensive earthquake preparedness planning guidelines, 1985. Available from: BAREPP, MetroCenter, 101 Eighth St, Suite 152, Oakland, CA 94607.

Bay Area Regional Earthquake Preparedness Project: County comprehensive earthquake preparedness planning guidelines, 1985. Available from: BAREPP, MetroCenter, 101 Eighth St, Suite 152, Oakland, CA 94607.

Bay Area Regional Earthquake Preparedness Project: Earthquake preparedness guidelines for hospitals, 1987. Available from: BAREPP, MetroCenter, Suite 152, 101 8th St, Oakland, CA 94607.

Bay Area Regional Earthquake Preparedness Project: Reducing the risks of nonstructural earthquake damage: a practical guide, ed 2, 1985. Available from: BAREPP, MetroCenter, Suite 152, 101 8th St, Oakland, CA 94607.

Baisden B: Social factors affecting mental health delivery: the case of disasters, in: Lewis EP, et al (eds), Sociological Research Symposium IX, Department of Sociology, Virginia Commonwealth University, Richmond, 1979, reprint no 125. Available from: Disaster Research Center, University of Delaware, Newark, 19716.

Baisden B, and Quarantelli EL: The delivery of mental health services in community disasters:

an outline of research findings, J Community Psychol 9(3):195, July 1981.

Baker GW, and Chapman DW (eds): Man and society in disaster, Basic Books, Inc, New York, 1962.

Bakst HJ, et al: The Worcester County tornado: a medical study of a disaster, National Academy of Sciences—National Research Council, Washington, DC, 1955.

Barton AH: Communities in disaster: a sociological analysis of collective stress situations, Doubleday, Garden City, NY, 1969.

Barton AH: Social organization under stress: a sociological review of disaster studies, disaster study no 17, publication no 1032, Disaster Research Group, National Academy of Sciences—National Research Council, Washington, DC, 1963.

Blanshan SA: Hospitals in "rough waters": the effects of a flood disaster on organizational change, (PhD dissertation), Department of Sociology, Ohio State University, 1975. Available from: University Microfilms International, 300 North Zeeb Rd, Ann Arbor, MI 48106.

Blanshan S, and Quarantelli EL: From dead body to person: the handling of fatal mass casualties in disasters, preliminary paper no 61, 1979. Available from: Disaster Research Center, University of Delaware, Newark, 19716.

Carroll JM: Computer simulation in emergency planning: Proceedings of the conference on computer simulation in emergency planning, January 27–29, 1983, San Diego, California, Simulation Series Vol 11, no 2. Available from: Society for Computer Simulation, PO Box 2228, La Jolla, CA 92038, $30.

Carroll JM: Emergency planning: proceedings of the conference on emergency planning, January 24–26, 1985, San Diego, California, Simulation Series Vol 15, no 1. Available from: Society for Computer Simulation, PO Box 2228, La Jolla, CA 92038, $30.

Chapman DW (ed): Human behavior in disaster: a new field of social research, (special issue), J Social Issues 10(3), 1954.

Clifford RA: Informal group actions in the Rio Grande flood, Committee on Disaster Studies, National Academy of Sciences—National Research Council, Washington, DC, 1955.

Clifford RA: The Rio Grande flood: a comparative study of border communities in disaster, Committee on Disaster Studies, National Research Council—National Academy of Sciences, Washington, DC, 1956.

Davenport S: Human response to hurricanes in Texas—two studies, Working Paper Series, no 34, 1978. Available from: Natural Hazards Research and Applications Information Center, University of Colorado, IBS no 6, Campus Box 482, Boulder, 80309, (303) 492-6818.

Demerath NJ, and Wallace AFC (eds): Human adaption to disaster, Human Organization, (special issue), 16(2), Summer 1957.

Drabek TE: Disaster in aisle 13: a case study of the coliseum explosion at the Indiana State Fairgrounds, October 31, 1963, 1968. Available from: Disaster Research Center, University of Delaware, Newark, 19716.

Drabek TE: Emergency management: the human factor, Federal Emergency Management Agency, 1985. Available from: Office of Programs and Academics, National Emergency Training Center, 16825 South Seton Avenue, Emmitsburg, MD 21727.

Drabek TE: Human system responses to disaster: an inventory of sociological findings, Springer-Verlag, New York, 1986. Available from: Springer-Verlag, PO Box 2485, Secaucus, NJ 07094.

Drabek TE: Taming the frontierland tornado: the emergent multiorganizational search and rescue network in Cheyenne, Wyoming, July, 1979, Tech Rep no 5, SAR Project, 1980. Available from: Department of Sociology, University of Denver, University Park, Denver, CO 80208.

Drabek TE: The professional emergency manager: Structures and strategy for success, monograph series no 44, 1987. Available from: Natural Hazards Information Center, IBS no 1, Campus Box 482, University of Colorado, Boulder, CO 80309.

Drabek TE, Tamminga HL, Kilijanek TS, et al: Managing multiorganizational emergency responses: emergent search and rescue networks in natural disaster and remote area settings, 1981. Available from: Natural Hazards Information Center, IBS no 6, Campus Box 482, University of Colorado, Boulder, CO 80309.

Dynes RR: Organized behavior in disaster, 1974. Available from: Disaster Research Center, University of Delaware, Newark, 19716.

Dynes RR, and Quarantelli EL: Organizational communications and decision making in crises, 1977. Available from: Disaster Research Center, University of Delaware, Newark, 19716.

Dynes RR, and Quarantelli EL: The role of local civil defense in disaster planning, 1977. Available from: Disaster Research Center, University of Delaware, Newark, 19716.

Dynes RR, Quarantelli EL, and Kreps GA: A perspective on disaster planning, ed 3, report series 11, 1981. Available from: Disaster Research Center, University of Delaware, Newark, 19716.

Federal Emergency Management Agency: Dam safety: an owner's guidance manual, FEMA 145/August 1987. Available from: FEMA Headquarters, 500 C St, SW, Washington, DC 20472.

Federal Emergency Management Agency: Emergency planning, IG-61, 1983. Available from: Publications Department, Federal Emergency Management Agency, PO Box 8181, Washington, DC 20024.

Federal Emergency Management Agency: Emergency planning, SM-61, 1983. Available from: Publications Department, Federal Emergency Management Agency, PO Box 8181, Washington, DC 20024.

Federal Emergency Management Agency: Emergency program manager: an orientation to the position, SS-1, 1983. Available from: Publications Department, Federal Emergency Management Agency, PO Box 8181, Washington, DC 20024.

Federal Emergency Management Agency: Exemplary practices in emergency management: business and industry council for emergency planning and preparedness, monograph series no 4, 1987. Available from: National Emergency Training Center, Emergency Management Institute, PO Box 70742, Washington, DC 20023.

Federal Emergency Management Agency: Exemplary practices in emergency management: North Dakota "Boys State" emergency simulation—a public-private experience, 1987. Available from: National Emergency Training Center, Emergency Management Institute, PO Box 70742, Washington, DC 20023.

Federal Emergency Management Agency: Exemplary practices in emergency management: San Mateo County, California, hazardous materials response plan and hazmat response unit, monograph series no 3, 1987. Available from: National Emergency Training Center, Emergency Management Institute, PO Box 70742, Washington, DC 20023.

Federal Emergency Management Agency: Exercise design course: exercise scenarios, SM-170.3, 1984. Available from: Publications Department, Federal Emergency Management Agency, PO Box 8181, Washington, DC 20024.

Federal Emergency Management Agency: Exercise design course: Guide to emergency management exercises, SM-170.2, 1984. Available from: Publications Department, Federal Emergency Management Agency, PO Box 8181, Washington, DC 20024.

Federal Emergency Management Agency: Exercise design course: IG-170, 1984. Available from: Publications Department, Federal Emergency Management Agency, PO Box 8181, Washington, DC 20024.

Federal Emergency Management Agency: Exercise design course: SM-170.1, 1984. Available from: Publications Department, Federal Emergency Management Agency, PO Box 8181, Washington, DC 20024.

Federal Emergency Management Agency: Formulating public policy in emergency management: course book and resource manual, SM-51, 1984. Available from: Publications Department, Federal Emergency Management Agency, PO Box 8181, Washington, DC 20024.

Federal Emergency Management Agency: Formulating public policy in emergency management, Instructor guide, 1983. Available from: Publications Department, Federal Emergency Management Agency, PO Box 8181, Washington, DC 20024.

Federal Emergency Management Agency: Guide to hurricane preparedness planning for state and local officials, CPG 2-16, 1984. Available from: Publications Department, Federal Emergency Management Agency, PO Box 8181, Washington, DC 20024.

Federal Emergency Management Agency: Hospital emergency department management of radiation emergencies, IG 80, 1984. Available from: Publications Department, Federal Emergency Management Agency, PO Box 8181, Washington, DC 20024.

Federal Emergency Management Agency: Hospital emergency department management of radiation accidents, SM 80, 1984. Available from: Publications Department, Federal Emergency Management Agency, PO Box 8181, Washington, DC 20024.

Federal Emergency Management Agency: Hurricane awareness workbook, no date. Available from: Publications Department, Federal Emergency Management Agency, PO Box 8181, Washington, DC 20024.

Federal Emergency Management Agency: Introduction to emergency management: instruction materials package, IG 60.1, 1983. Available from: Publications Department, Federal Emergency Management Agency, PO Box 8181, Washington, DC 20024.

Federal Emergency Management Agency: Introduction to emergency management: SM-60, 1983. Available from: Publications Department, Federal Emergency Management Agency, PO Box 8181, Washington, DC 20024.

Federal Emergency Management Agency: Job aid manual, 1983. Available from: Publications Department, Federal Emergency Management Agency, PO Box 8181, Washington, DC 20024.

Federal Emergency Management Agency: Perspectives on hurricane preparedness: techniques in use today, 1984. Available from: FEMA, 500 C Street, SW, Washington, DC 20472.

Faupel CE: The ecology of disaster, 1985. Available from: Irvington Publishers, Inc, 740 Broadway, New York, NY 10003.

Form W, and Nosow S: Community in disaster, Harper, New York, 1958.

Form WH, Nosow S, Stone GP, et al: Final report on the Flint-Beecher tornado, (unpublished report), Social Research Service, Continuing Education Service, Michigan State College, 1954.

Fritz CE: Disaster, in: Merton RK, Nisbet RA (eds), Contemporary social problems, Harcourt, Brace, & World, New York, 1961.

Fritz C, and Marks E: The NORC studies of human behavior in disaster, J Social Issues 10(3):26, 1954.

Fritz CE, and Mathewson JH: Convergence behavior in disasters: a problem in social control, Disaster Study no 9, publication no 476, Committee on Disaster Studies, National Academy of Sciences—National Research Council, Washington, DC, 1956.

Kennedy WC: Some preliminary observations on a hospital response to the Jackson, Mississippi tornado of March 3, 1966, research report 17, 1967. Available from: Disaster Research Center, University of Delaware, Newark, 19716.

Kennedy WC, Brooks MJ, and Vargo SM: The police department in natural disaster operations, 1969. Available from: Disaster Research Center, University of Delaware, Newark, 19716.

Kilijanek TS: The night of the Whippoorwill: the search and rescue response to a boating disaster, Tech Rep no 2, SAR Research Project, 1980. Available from: Department of Sociology, University of Denver, University Park, Denver, CO 80208.

Kilijanek TS: There she blows: the search and rescue response to the Mount St. Helens volcano, Tech Rep no 11, SAR Research Project, 1981. Available from: Department of Sociology, University of Denver, University Park, Denver, CO 80208.

Koegler RR, and Hicks SM: The destruction of a medical center by earthquake: initial effects on patients and staff, Western J Med 116:63, Feb 1972.

Lantis M: When the earthquake hits home: Anchorage in the "Great Alaska Earthquake," 1984. Available from: Natural Hazards Research and Applications Information Center, IBS no 6, Campus Box 482, University of Colorado, Boulder, CO 80309.

Lavalla R, and Stoffel S: Blueprint for community emergency management: a text for managing emergency operations, 1983. Available from: Emergency Response Institute, 1819

Mark St, NE, Olympia, WA 98506, (206) 491-7785 or (509) 782-4832.

Lavalla R, Stoffel S, and Kartez J: Community emergency management: development and strategies. Available from: Emergency Response Institute, 1819 Mark St, NE, Olympia, WA 98506, (206) 491-7785 or (509) 782-4832.

Marks ES, Fritz CE, et al: Human reactions in disaster situations, volumes 1–3, (unpublished report), National Opinion Research Center, University of Chicago, June 1954. Available to qualified Armed Services Technical Information Agency users as ASTIA document no AD-107 594.

Mileti DS, Drabek TE, and Haas JE: Human systems in extreme environments: a sociological perspective, monograph no 21, Program on Technology, Environment and Man, 1975. Available from: Natural Hazards Information Center, IBS no 6, Campus Box 482, University of Colorado, Boulder, CO 80309.

Moore HE: And the winds blew, Hogg Foundation for Mental Health, University of Texas, 1964.

Moore HE: Before the wind, National Academy of Sciences—National Research Council, Washington, DC, 1963.

Moore HE: Tornados over Texas: a study of Waco and San Angelo in disaster, University of Texas Press, Austin, 1958.

National Council on Radiation Protection and Measurements: Management of persons accidentally contaminated with radionucleotides, NCRP report no 65. Available from: The Council at 7910 Woodmont Avenue, Washington, DC 20014.

Office of the United Nations Disaster Relief Coordinator: Disaster prevention and mitigation: a compendium of current knowledge, vol 11: Preparedness aspects, 1984. Available from: UNDRO Liaison Office, Office of the Disaster Relief Co-ordinator, United Nations Secretariat Bldg, First Avenue and 42nd St, Rm S-2935, New York, NY 10017.

Office of the United Nations Disaster Relief Coordinator: Disaster prevention and mitigation: a compendium of current knowledge, vol 12: Social and sociological aspects, 1986. Available from: UNDRO Liaison Office, Office of the Disaster Relief Co-ordinator, United Nations Secretariat Bldg, First Avenue and 42nd St, Rm S-2935, New York, NY 10017.

Pan American Health Organization: A guide to emergency health management after natural disaster, scientific publication no 407, 1981. Available from: Pan American Health Organization, Pan American Sanitary Bureau, Regional Office of the World Health Organization, 525 23rd St, NW, Washington, DC 20037.

Petak WJ: Emergency management: a challenge for public administration (special issue), Publ Admin Rev 45, Jan 1985. Available from: American Society for Public Administration, 1120 G St, NW, Suite 500, Washington, DC 20005, (202) 393-7878.

Quarantelli EL: An assessment of conflicting views on the consequences of community disasters for mental health, preliminary paper no 89, 1984. Available from: Disaster Research Center, University of Delaware, Newark, 19716.

Quarantelli EL: An assessment of conflicting views on mental health: the consequences of traumatic events, 1985. Available from: Disaster Research Center, University of Delaware, Newark, 19716.

Quarantelli EL: An overview of research of PTSD in survivors of disasters, working paper no 72, 1984. Available from: Disaster Research Center, University of Delaware, Newark, 19716.

Quarantelli EL: Delivery of emergency medical care in disasters: assumptions and realities, 1983. Available from: Irvington Publishers, Inc, 551 Fifth Ave, New York, NY 10017.

Quarantelli EL: Disasters: theory and research, 1978. Available from: Sage Publications, 275 South Beverly Dr, Beverly Hills, CA 90212.

Quarantelli EL: Images of withdrawal behavior in disasters: some basic misconceptions, Soc Problems 8(1):68, Summer 1960.

Quarantelli EL: Mass behavior and governmental breakdown in major disasters: viewpoint of a researcher, Police Yearbook, p 105, 1965.

Quarantelli EL: Organizational behavior in disasters and implications for disaster planning, report series 18, 1985. Available from: Disaster Research Center, University of Delaware, Newark, 19716.

Quarantelli EL: Sociobehavioral responses to chemical hazards: preparations for and responses to acute chemical emergencies at the local community level, 1981. Available from: Disaster Research Center, University of Delaware, Newark, 19716.

Quarantelli EL: The community general hospital: its immediate problems in disasters, Am Behav Sci 13(3):389, Jan–Feb 1970.

Quarantelli EL: The consequences of disasters for mental health: conflicting views, preliminary paper no 62, 1979. Available from: Disaster Research Center, University of Delaware, Newark, 19716.

Quarantelli EL: The delivery of emergency medical services (EMS) in disasters: recommendations from field studies, preliminary paper no 67, 1981. Available from: Disaster Research Center, University of Delaware, Newark, 19716.

Quarantelli EL: The reality of local community chemical disaster preparedness: three case studies, 1981. Available from: Disaster Research Center, University of Delaware, Newark, 19716.

Quarantelli EL, and Dynes RR (eds): Organizational and group behavior in disaster, (special issue), Am Behav Sci 13(3), Jan/Feb 1970.

Quarantelli EL, and Dynes RR: When disaster strikes (It isn't much like what you've heard or read about), Psychology Today, p 67, Feb 1972.

Raker JW, Wallace AFC, and Rayner JF: Emergency medical care in disasters: a summary of recorded experience, disaster study no 6, publication no 457, Disaster Research Group, National Academy of Sciences—National Research Council, Washington, DC, 1956.

Reynolds S, and Wright JE: A selective literature review of disaster medical services, working paper 64, 1976. Available from: Disaster Research Center, University of Delaware, Newark, 19716.

Seismic Safety Commission: Preliminary reports submitted to the Seismic Safety Commission on the May 2, 1983 Coalinga earthquake, publication no SSC 83-08, 1983. Available from:

The Commission, 1900 K St, Suite 100, Sacramento, CA 95814.

Seismic Safety Commission: Public official attitudes toward disaster preparedness in California, publication no SSC 79-05, 1979. Available from: The Commission, 1900 K St, Suite 100, Sacramento, CA 95814.

Shore JH: Disaster stress studies: new methods and findings, 1986. Available from: American Psychiatric Press, Inc, Department 99, 1400 K St, NW, Washington, DC 20005, (800) 368-5777.

Smith D (ed): Disasters and disaster relief, (special issue), Ann Amer Acad Polit Soc Sci:309, Jan 1957.

Sorensen JH: Emergency response to Mount St. Helens' eruption: March 20 to April 10, 1980, working paper series no 43, 1981. Available from: Natural Hazards Research and Applications Information Center, University of Colorado, IBS no 6, Campus Box 482, Boulder, CO 80309, (303) 492-6818.

Sorensen JH, Mileti DS, and Copenhaver E: Inter and intraorganizational cohesion in emergencies, Mass Emergencies and Disasters 3(3):27, Nov 1985.

Tamminga HL: Search and rescue operations following the Texas Hill Country flash floods, Tech Rep no 3, SAR Research Project, 1981. Available from: Department of Sociology, University of Denver, University Park, Denver, CO 80208.

Taylor V (ed): Delivery of emergency medical services in disasters, (special issue), Mass Emerg 2(3), Sept 1977.

Tierney KJ: A primer for preparedness for acute chemical emergencies, book and monograph series no 14, 1980. Available from: The Disaster Research Center, University of Delaware, Newark, 19716.

Tierney KJ: Report on the Coalinga earthquake of May 2, 1983, publication no SSC 85-01, 1985. Available from: Seismic Safety Commission, 1900 K St, Suite 100, Sacramento, CA 95814.

Wallace AFC: An exploratory study of individual and community behavior in an extreme situation: Tornado in Worcester, Disaster Study no 3, publication no 392, Committee on Disaster Studies, National Academy of Sciences—

National Research Council, Washington, DC, 1956.

Wallace AFC: Human behavior in extreme situations: a survey of the literature and suggestions for further research, report no 1, Committee on Disaster Studies, National Academy of Sciences—National Research Council, Washington, DC, 1956.

Washington State University, Environmental Research Center: Adaptive planning for community emergency management: a summary for public managers, 1987. Available from: The Center, Pullman, WA 99164-4430.

Wenger DE, Dykes JD, and Sebok TD: It's a matter of myths: an empirical examination of individual insight into disaster response, Mass Emer 1(1):33, Oct 1975.

Wenger DE, James TF, and Faupel CE: Disaster beliefs and emergency planning, 1985. Avail-able from: Irvington Publishers, Inc, 740 Broadway, New York, NY 10003.

Wenger D, Quarantelli EL, and Dynes R: Disaster analysis: emergency management offices and arrangements, final project report no 34, 1986. Available from: Disaster Research Center, University of Delaware, Newark, 19716.

Williams HB, and Rayner JF: Emergency medical services in disaster, Med Ann District of Columbia 25(12):655, Dec 1956.

Wright JE: Interorganizational relations as structure and as action: the case for emergency medical services in disaster, 1977. Available from: The Disaster Research Center, University of Delaware, Newark, 19716.

Yutzy D: Community priorities in the Anchorage, Alaska earthquake, 1964, Disaster Research Center monograph series no 4, 1969. Available from: The Disaster Research Center, University of Delaware, Newark, 19716.

REFERENCES

(ACEP)
American College of Emergency Physicians, Disaster Committee: Student manual for disaster management and planning for emergency physician's course, Federal Emergency Management Agency, Emergency Management Institute, Emmitsburg, MD, no date.

(ACEP, 1976)
American College of Emergency Physicians: The role of the emergency physician in mass casualty/disaster management: ACEP position paper, J Am Coll Emerg Phys 5(11):901, 1976.

(ACS, 1986)
Caring for the injured patient (special issue), Bull Am Coll Surg, 71(10), 1986.

(Adams, 1980)
Adams CR et al: The organization of search and rescue efforts following the Wichita Falls, Texas, tornado. A paper presented at the annual meeting of the American Sociological Association, New York, August 1980.

(Adams, 1981a)
Adams CA: Law: its effects on interorganizational authority structures in post-disaster responses, Tech Rep no 9, SAR Project, Department of Sociology, University of Denver, Denver, CO, 1981.

(Adams, 1981b)
Adams CR: Search and rescue efforts following the Wichita Falls tornado, Tech Rep no 4, SAR Project, Department of Sociology, University of Denver, Denver, CO, 1981.

(Adams, 1982)
Adams R: DC disaster crisis: a fatal plane crash and deadly subway derailment challenge regional emergency forces, Firehouse, p 50, March 1982.

(Anderson, 1965)
Anderson W: Some observations on a disaster subculture: the organizational response of Cincinnati, Ohio, to the 1964 flood, research note no 6, Disaster Research Center, University of Delaware, Newark, 1965.

(Arnett, 1981)
Arnett N: The Las Vegas jinx, Emergency Product News, p 76, May 1981.

(Arnett, 1983)
Arnett N: The Coalinga earthquake: an EMS insight, Emergency 15:30, July 1983.

(Bahme, 1978)
Bahme CW: Fire officer's guide to disaster control, National Fire Protection Association, Quincy, MA, 1978.

(Baker, 1962)
Baker CW, and Chapman DW, editors: Man and society in disaster, New York, 1962, Basic Books Inc.

(Baker, 1979)
Baker FJ: The management of mass casualty disasters, Topics in Emerg Med 1(1):149, May 1979.

(Barton, 1963)
Barton AH: Social organization under stress: a sociological review of disaster studies, disaster study no 17, publication no 1032, Disaster Research Group, National Academy of Sciences—National Research Council, Washington, DC, 1963.

(Barton, 1969)
Barton A: Communities in disaster: a sociological analysis of collective stress situations, Garden City, NY, 1969, Doubleday & Co.

(Bates, 1963)
Bates FL, et al: The social and psychological consequences of a natural disaster: a longitudinal study of Hurricane Audrey, study no 18, Disaster Research Group, National Academy of Sciences—National Research Council, Washington, DC, 1963.

(Berke, 1987)
Berke P: Hurricane hazard mitigation: how are we doing?, Natural Hazards Observer 11(5):1, 1987.

(Bernstein, 1986)
Bernstein AB: The emergency public relations manual, revised, Highland Park, NJ, 1986, PASE Inc.

(Best, 1980)
Best R: The evacuation of Mississauga: Part II, National Fire Protection Association, Quincy, MA, 1980.

(Best, 1982)
Best R, and Demers DP: Investigation report on the MGM Grand Hotel fire, Las Vegas, Nevada, November 21, 1980, National Fire Protection Association, Quincy, MA, 1982.

(Blanchard, 1985)
Blanchard BW: American civil defense 1945–1984: the evolution of programs and policies, Monograph Series, Federal Emergency Management Agency, National Emergency Training Center, Emmitsburg, MD, 1985.

(Blanshan, 1978)
Blanshan SA: A time model: hospital organizational response to disaster. In Quarantelli EL: Disasters: theory and research, Beverly Hills, CA, 1978, Sage Publications.

(Bolduc, 1987)
Bolduc JP: Natural disasters in developing countries: myths and the role of the media, Emerg Prepared Digest (Canada) 14(3):12, July–Sept 1987.

(Bourque, 1973)
Bourque LB, Reeder LG, and Cherlin A: The unpredictable disaster in a metropolis: public response to the Los Angeles earthquake of February 1971, report #SRC-TR1-73 (AD-7655130), (microfilm), University of California at Los Angeles, Survey Research Center, 1973.

(Bowers, 1960)
Bowers WF, et al: Surgical philosophy in mass casualty management, Springfield, IL, 1960, CC Thomas.

(Boyd, 1985)
Boyd J: TIS radio keeps the public informed, Law and Order 33(11):63, 1985.

(Bronson, 1959)
Bronson W: The earth shook, the sky burned, Garden City, NY, 1959, Doubleday & Co.

(Brunacini, 1978)
Brunacini AV: Phoenix Fire Department operations manual, vol 2, standard operating procedures, unpublished, Phoenix, AZ, 1978.

(Brunacini, 1985)
Brunacini AV: Fire command, National Fire Protection Association, Quincy, MA, 1985.

(Bryan, 1982)
Bryan JL: Human behavior in the MGM Grand Hotel fire, Fire J 76:37, March 1982.

(Buerk, 1982)
Buerk CA, et al: The MGM Grand Hotel fire: lessons learned from a major disaster, Arch Surg 117(5):641, 1982.

(Burkle, 1984)
Burkle FM Jr, Sanner PH, and Wolcott BW: Disaster medicine: application for the immediate management and triage of civilian and military disaster victims, New Hyde Park, NY, 1984, Medical Examination Publishing Co, Inc.

(Burton, 1968)
Burton I, Kates RW, and White GF: The human ecology of extreme geophysical events, working paper series no 1, Natural Hazards Research and Applications Information Center, University of Colorado, Boulder, 1968.

(Bush, 1981)
Bush S: Disaster planning and multiagency coordination (unpublished paper), Emergency Planning Department, City of Littleton, CO, 1981.

(Butman, 1982)
Butman AM: Responding to the mass casualty incident: a guide for EMS personnel, Emergency Training, Division of Educational Direction, Inc, Westport, CT, 1981.

(CFCA, 1986)
California Fire Chief's Association: Multi-casualty incident operational procedures manual, Rio Linda, CA, 1986.

(CSTI, 1987)
California Specialized Training Institute: The public information officer in disasters and haz mat incidents, a seminar/video workshop, held at the "Cliff House," University of California at Santa Barbara, August 26–27, 1987.

(Carroll, 1983)
Carroll JM: Computer simulation in emergency planning: proceedings of the conference on computer simulation in emergency planning, January 27–29, 1983, San Diego, CA, simulation series vol 11, no 2, Society for Computer Simulation, La Jolla, CA, 1983.

(Carroll, 1985)
Carroll JM: Emergency planning: proceedings of the conference on emergency planning, January 24–26, 1985, San Diego, CA, simulation series vol 15, no 1, Society for Computer Simulation, La Jolla, CA, 1985.

(Casper, 1983)
Casper AC, and Herman RE: Emergency preparedness—everybody must be involved, Fire Engineering 136(7):39, 1983.

(Chase, 1980)
Chase RA: FIRESCOPE: a new concept in multiagency fire suppression coordination, general technical report PSW-40, US Department of Agriculture, Forest Service, Pacific Southwest Forest and Range Experiment Station, Berkeley, CA, 1980.

(Cigler, 1986)
Cigler B: Emergency management and public administration, report 86–12, University of North Carolina at Chapel Hill, 1986.

(Cihlar, 1972)
Cihlar C: Hospitals respond efficiently to Chicago's worst train wreck, Hospitals-JAHA 46:17, Nov 16, 1972.

(Clary, 1985)
Clary BB: The evolution and structure of natural hazard policies. In Petak WJ: Emergency management: a challenge for public administration, special issue, Public Admin Rev 45:20, Jan 1985.

(Cohen, 1982a)
Cohen E: Disasters: an emergency care workbook, Instructional Development and Educational Aids, Inc, San Diego, CA, 1982.

(Cohen, 1982b)
Cohen E: Triage 1982: you've trained for "the big one," but are you ready for the little disaster, J Emerg Med Serv, p 23, Aug 1982.

(Cohen: 1983)
Cohen E: A better mousetrap: what makes up the "perfect" triage tag?, J Emerg Med Serv, p 30, July 1983.

(Cohen, 1986)
Cohen E: Patient identification: a look at triage tags, J Emerg Med Serv 15(9):45, Oct 1986.

(Coleman, 1978)
Coleman RJ: Management of fire service operations, North Scituate, MA, 1978, Duxbury Press.

(Comm on Disasters, 1980)
Committee on Disasters and the Mass Media, Commission on Sociotechnical Systems: Disasters and the mass media, National Research Council—National Academy of Sciences, Washington, DC, 1980.

(Committee on EMS, 1971)
Committee on Emergency Medical Services of the Medical Association of Georgia, the Georgia State Committee on Trauma of the American

College of Surgeons, the Disaster Planning Committee of the Georgia Hospital Association: The mass casualty/disaster plan, reprint, Bull Am Coll Surg 56:9, Nov 1971.

(Cowley, 1982)
Cowley RA: Mass casualty needs. In Cowley RA: Mass casualties: a lessons learned approach, proceedings: First international assembly on emergency medical services, Baltimore, June 13–17, 1982, DOT HS 806 302, US Department of Transportation, National Highway Traffic Safety Administration, Washington, DC, 1982.

(Cowley, 1985)
Cowley RA: The use of communications satellite systems in major disaster situations. In Manni C and Magalini SI: Emergency and disaster medicine, New York, 1985, Springer-Verlag.

(Davenport, 1978)
Davenport S: Human response to hurricanes in Texas—two studies, working paper series no 34, Natural Hazards Research and Applications Information Center, University of Colorado, Boulder, 1978.

(DeAtley, 1982)
DeAtley CA: One hundred fifty minutes: a double tragedy in the nation's capital, J Emerg Med Serv 7(3):26, 1982.

(Deats, 1985)
Deats RV: Tornado touch down, Fire Command 52(5):18, 52, 1985.

(Dektar, 1971)
Dektar C: Extensive rescue operations follow Los Angeles quake, Fire Engineering, 124:38–40, June 1971.

(Dick, 1982)
Dick T, et al: 1982 almanac of emergency medical services, J Emerg Med Serv 7(1):41, 1982.

(Dickson, 1987)
Dickson R: Press relations: an automated news media system, FBI Law Enforcement Bull, p 18, Feb 1987.

(Dinerman, 1988)
Dinerman N, Personal communication, Medical Director, Denver Municipal Ambulance, Faculty Physician, Denver General Hospital Emergency Department, 1988.

(Div Med Sci, 1966)
Division of Medical Sciences, National Academy of Science, National Research Council: Accidental death and disability: the neglected disease of modern society, US Dept Health, Education, and Welfare, Public Health Service, Health Services and Mental Health Admin, Div of Emergency Health Services, Rockville, MD, 1966.

(Dorn, 1986)
Dorn JM, and Hopkins BM: The handling, decontamination, and transportation of radiation fatalities. In: Multiple death disaster response workshop, SM/IG 193, Federal Emergency Management Agency and the National Funeral Directors Association, Washington, DC, 1986.

(Drabek, 1968)
Drabek TE: Disaster in aisle 13: a case study of the coliseum explosion at the Indiana State fairgrounds, October 31, 1963, Disaster Research Center, University of Delaware, Newark, 1968.

(Drabek, 1980)
Drabek TE: Taming the frontierland tornado: the emergent multiorganizational search and rescue network in Cheyenne, WY, July 1979, Tech Rep no 5, SAR Project, Department of Sociology, University of Denver, CO, 1980.

(Drabek, 1981)
Drabek TE, et al: Managing multiorganizational emergency responses: emergent search and rescue networks in natural disaster and remote area settings, Natural Hazards Information Center, University of Colorado, Boulder, 1981.

(Drabek, 1985a)
Drabek TE: Managing the emergency response. In Petak WJ: Emergency management: a challenge for public administration (special issue), Public Admin Rev 45:85, Jan 1985.

(Drabek, 1985b)
Drabek TE: Emergency management: the human factor, Federal Emergency Management Agency, National Emergency Training Center, Emmitsburg, MD, 1985.

(Drabek, 1986)
Drabek TE: Human system responses to disaster: an inventory of sociological findings, New York, 1986, Springer-Verlag.

(Drabek, 1987)
Drabek TE: The professional emergency manager: structures and strategy for success, mono-

graph no 44, Institute of Behavioral Science, University of Colorado, Boulder, 1987.

(Dunlap, 1981)
Dunlap MJ: Nurses assist injured at Hyatt disaster, Am Nurse 13(8)1,7,10, Sept 1981.

(Dynes, 1968)
Dynes RR, and Quarantelli EL: What looting in civil disturbances really means, Trans-Action 5:9, May 1968.

(Dynes, 1970)
Dynes RR: Organizational involvement and changes in community structure in disaster, Am Behav Sci 13(3)430, Jan/Feb 1970.

(Dynes, 1974)
Dynes RR: Organized behavior in disaster, Disaster Research Center, University of Delaware, Newark, 1974.

(Dynes, 1977)
Dynes RR, and Quarantelli EL: Organizational communications and decision making in crises, Disaster Research Center, University of Delaware, Newark, 1977.

(Dynes, 1978)
Dynes RR: Interorganizatonal relations in communities under stress. In Quarantelli EL: Disasters: theory and research, Beverly Hills, CA, 1978, Sage Publications.

(Dynes, 1981)
Dynes RR, Quarantelli EL, and Kreps GA: A perspective on disaster planning, ed 3, report series 11, Disaster Research Center, University of Delaware, Newark, 1981.

(Edelstein, 1982)
Edelstein S: Metro subway accident. In Cowley RA: Mass casualties: a lessons learned approach, Proceedings of the first international assembly on emergency medical services, Maryland Institute for Emergency Medical Services, Baltimore, MD, June 13–17, 1982, US Department of Transportation, National Highway Traffic Safety Administration, Washington, DC, 1982.

(Esch, 1982)
Esch VH: Air crash on the Potomac: onsite problems. In Cowley RA: Mass casualties, a lessons learned approach, Proceedings of the first international assembly on emergency medical services, Maryland Institute for Emergency Medical Services, Baltimore, MD, June 13–17 1982,

US Department of Transportation, National Highway Traffic Safety Administration, Washington, DC, 1982.

(FEMA)
Federal Emergency Management Agency: Hurricane awareness workbook, Washington, DC, no date.

(FEMA, 1981)
Federal Emergency Management Agency: Disaster operations: a handbook for local governments, Washington, DC, 1981.

(FEMA, 1983a)
Federal Emergency Management Agency: Emergency program manager: an orientation to the position, SS-1, Washington, DC, 1981.

(FEMA, 1983b)
Federal Emergency Management Agency: Introduction to emergency management, SM-60, Washington, DC, 1983.

(FEMA, 1983c)
Federal Emergency Management Agency: Emergency planning: preclass activities, Washington, DC, 1983.

(FEMA, 1983d)
Federal Emergency Management Agency: Emergency planning, SM-61, Washington, DC, 1983.

(FEMA, 1983e)
Federal Emergency Management Agency: Job aid manual, National Emergency Training Center, Emergency Management Institute, Emmitsburg, MD, 1983.

(FEMA, 1984a)
Federal Emergency Management Agency: Formulating public policy in emergency management: course book and resource manual, SM-51, Washington, DC, 1984.

(FEMA, 1984b)
Federal Emergency Management Agency: Emergency operating centers handbook, CPG 1-20, Washington, DC, 1984.

(FEMA, 1984c)
Federal Emergency Management Agency: Objectives for local emergency management, CPG 1-5, Washington, DC, 1984.

(FEMA, 1985a)
Federal Emergency Management Agency: Integrated emergency management system: hazard identification, capability assessment, and

multi-year development plan—overview, CPG 1-34, Washington, DC, 1985.

(FEMA, 1985b)
Federal Emergency Management Agency: Integrated emergency management system: hazard identification, capability assessment and multi-year development plan for local governments—workbook, CPG 1-35, Washington, DC, 1985.

(FEMA, 1985c)
Federal Emergency Management Agency: Integrated emergency management system: hazard identification, capability assessment, and multi-year development plan for local governments—response book, CPG-135a, Washington, DC, 1985.

(FEMA, 1985d)
Federal Emergency Management Agency: Guide for development of state and local emergency operations plans: interim guidance, CPG 1-8, Washington, DC, 1985.

(FEMA, 1987)
Federal Emergency Management Agency: Exemplary practices in emergency management: The California FIRESCOPE Program, monograph series no 1, National Emergency Training Center, Emergency Management Institute, Washington, DC, 1987.

(Faupel, 1985)
Faupel CE: The ecology of disaster, New York, 1985, Irvington Publishers, Inc.

(Form, 1958)
Form W, and Nosow S: Community in disaster, New York, 1958, Harper.

(Foster, 1980)
Foster HD: Disaster planning: the preservation of life and property, New York, 1980, Springer-Verlag.

(Fox, 1981)
Fox GG: Disaster planning: a tale of two cities, Public Management 63:7, Jan/Feb 1981.

(Fritz, 1954)
Fritz C, and Marks E: The NORC studies of human behavior in disaster, J Social Issues, 10(3):26, 1954.

(Fritz, 1956)
Fritz CE, and Mathewson JH: Convergence behavior in disasters: a problem in social control, disaster study no 9, publicaton no 476, Committee on Disaster Studies, National Academy of Sciences—National Research Council, Washington, DC, 1956.

(Fritz, 1957)
Fritz CE, and Williams HB: The human being in disasters: a research perspective, Ann Amer Acad of Political and Social Sciences 309:42–51, Jan 1957.

(Fritz, 1961)
Fritz CE: Disaster. In Merton RK, and Nisbet RA, editors: Contemporary social problems, New York, 1961, Harcourt, Brace, & World.

(Gallagher, 1985)
Gallagher J, and Gallagher D: Delta, go around: a trial by fire for rescuers, J Emerg Med Serv, p 42, Oct 1985.

(Gann, 1979)
Gann DS, et al: Mass casualty management. In Zuidema GD, Rutherford RB, Ballinger WF: The management of trauma, ed 3, Philadelphia, 1979, WB Saunders.

(Gazzaniga, 1979)
Gazzaniga AB, Iseri LT, and Baren M: Emergency care: principles and practices for the EMT-paramedic, Reston, VA, 1979, Reston Publishing Co.

(Gibson, 1977)
Gibson G: Disaster and emergency medical care: methods, theories and a research agenda, Mass Emerg 2:195, 1977.

(Glass, 1980)
Glass RI, et al: Injuries from the Wichita Falls tornado: implications for prevention, Science 207:734, Feb 15 1980.

(Golec, 1977)
Golec JA, and Gurney PJ: The problem of needs assessment in the delivery of EMS, Mass Emerg 2:169, 1977.

(Goodwin, 1982)
Goodwin DV: DC crash problems magnified by snow, traffic snarl and EMS snafu, Emerg Dept News 4(2):1,14, Feb 1982.

(Gordon, 1986)
Gordon D: High-rise fire rescue: lesson from Las Vegas, Emerg Med Serv 15(5):20, June 1986.

(Grant, 1982)
Grant HD, Murray RH Jr, and Bergeron JD: Emergency care, ed 3, Bowie, MD, 1982, Brady Communications Co.

(Gratz, 1972)
Gratz DB: Fire department management: scope and method, Fire Science Series, Beverly Hills, CA, 1972, Glencoe Press.

(Gray, 1981)
Gray C, and Knabe H: The night the skywalks fell: 111 dead, 188 injured in Kansas City hotel collapse, Firehouse 6:66,132, Sept 1981.

(Grolier, 1985)
Encyclopedia Americana, vol 9, Danbury CT, 1985, Grolier, Inc.

(Grollmes, 1985)
Grollmes EE: Air disaster response planning: lessons for the future, Federal Emergency Management Agency, National Emergency Training Center, Emmitsburg, MD, 1985.

(Hamilton, 1955)
Hamilton RV, Taylor RM, and Rice GE Jr: A social psychological interpretation of the Udall, Kansas, tornado, unpublished report, University of Wichita, Oct 1985.

(Harter, 1985)
Harter S: Emergency broadcast system in California, unpublished, State of California, Office of Emergency Services, Telecommunications Division, Sacramento, 1985.

(Hartsough, 1985)
Hartsough DM, and Mileti DS: The media in disaster. In Laube J, and Murphy SA: Perspectives on disaster recovery, Norwalk, CT, 1985, Appleton-Century-Crofts.

(Hildebrand, 1980)
Hildebrand MS: Disaster planning guidelines for fire chiefs, International Association of Fire Chiefs, Inc, publication AD AO91, Department of Commerce, National Technical Information Service, Springfield, VA, 1980.

(Holloway, 1978)
Holloway RD, Steliga JF, and Ryan CT: The EMS system and disaster planning: some observations, J Am Coll Emerg Phys 7(2):60, 1978.

(Holton, 1985)
Holton JL: The electronic media and disasters in the high-tech age, Federal Emergency Management Agency, National Emergency Training Center, Emmitsburg, MD, 1985.

(Houghton, 1986)
The 1986 information please almanac, Boston, 1986, Houghton Mifflin Co.

(ICS, 1980)
FIRESCOPE Program: Multi-agency coordination system (MACS): radio communication guidelines, FP 441-1, Operations Coordination Center, FIRESCOPE Program, Riverside, CA, 1980.

(ICS, 1981)
FIRESCOPE Program: Incident Command System position manual: information officer, ICS-220-2, Operations Coordination Center, FIRESCOPE Program, Riverside, CA, 1981.

(ICS, 1982)
National Interagency Incident Management System (NIIMS): Basic ICS Course, instructor's manual, I-220, US Department of the Interior, Bureau of Land Management, National Wildfire Coordinating Group, Boise Interagency Fire Center, Boise, ID, 1982.

(ICS, 1983)
National Interagency Incident Management System: Information and guides, Publication Management System, National Wildfire Coordination Group, Boise Interagency Fire Center, Boise, ID, 1983.

(ICS, 1985a)
FIRESCOPE Program: Incident Command System: incident commander: instructor's guide, I-400, Operations Coordination Center, FIRESCOPE Program, Riverside, CA, 1985.

(ICS, 1985b)
FIRESCOPE Program: Planning section chief: instructor's guide, I-440, Operations Coordination Center, FIRESCOPE Program, Riverside, CA, 1985.

(ICS, 1986)
OES Fire and Rescue Service Advisory Committee, FIRESCOPE Board of Directors: FIRESCOPE decision process, MACS 410-4, Operations Coordination Center, FIRESCOPE Program, Riverside, CA, 1986.

(ICS, 1987)
MACS procedures guide, MACS 410-1, multi-agency coordination system, Operations Coordination Center, FIRESCOPE Program, Riverside, CA, 1987.

(Irwin, 1980)
Irwin RL: Unified command, unpublished training document, FIRESCOPE program, US Forest Service, Riverside, CA, 1980.

(Irwin, 1984)
Irwin RL: Riverside County elected and appointed officials training seminar (unpublished), Riverside, CA, 1984.

(Irwin, 1987)
Irwin RL: Personal communication, Emergency Management Consultant, Basic Governmental Services, Sonora, CA, 1987.

(Irwin, 1988)
Irwin RL: Personal communication, Emergency Management Consultant, Basic Governmental Services, Sonora, CA, 1988.

(Irwin)
Irwin RL: Inyo-Mono seminars: the Incident Command System, unpublished, no date.

(Jaworski, 1954)
Jaworski H: The Waco tornado, Bull Am Coll Surg 39:129, May/June 1954.

(JCFEDS, 1983)
Joint Committee on Fire, Police, Emergency and Disaster Services: California's emergency communications crises, California State Senate and Assembly, April 1983.

(Jenkins, 1975)
Jenkins AL: Emergency department organization and management, St Louis, 1975, The C V Mosby Co.

(Johnson, 1986)
Johnson B: Dealing with mass media in time of emergency. In: Multiple death disaster response workshop, Federal Emergency Management Agency and the National Funeral Directors Association, Washington, DC, 1986.

(KC Health Department, 1981)
Health Department, Kansas City, MO: Hyatt disaster medical assessment, unpublished, 1981.

(Kallsen, 1983)
Kallsen G: Collapse of Coalinga, J Emerg Med Serv 8(7):24, 1983.

(Kasperson, 1985)
Kasperson RE, and Pijawka DK: Societal response to hazards and major hazard events: comparing natural and technological hazards. In Petak WJ: Emergency management: a challenge for public administration (special issue), Public Admin Rev 45:7, Jan 1985.

(Kast, 1974)
Kast FE, and Rozensweig JE: Organization and management: a systems approach, New York 1974, McGraw-Hill, Inc.

(Keating, 1982)
Keating JP: The myth of panic. In: Hotel fires, behind the headlines, pp 89, 107, National Fire Protection Association, Quincy, MA, 1982.

(Kennedy, 1970)
Kennedy WC: Police departments: organization and tasks in disaster, Am Behav Sci 13(3):354, Jan–Feb 1970.

(Kilijanek, 1979)
Kilijanek TS, et al: The emergence of a post-disaster communications network, a paper presented at the Annual Meeting of the American Sociological Association, Boston, MA, Aug 1979.

(Kilijanek, 1980)
Kilijanek TS: The night of the Whippoorwill: the search and rescue response to a boating disaster, Tech Rep no 2, SAR Research Project, Department of Sociology, University of Denver, CO, 1980.

(Kilijanek, 1981)
Kilijanek TS: There she blows: the search and rescue response to the Mount St Helens volcano, Tech Rep no 11, SAR Project, Department of Sociology, University of Denver, CO, 1981.

(Killian, 1953)
Killian LM, and Rayner JF: Assessment of disaster operations following the Warner Robins tornado, Committee on Disaster Studies, National Academy of Sciences—National Research Council, Washington, DC, 1953.

(Koehler, 1987)
Koehler G: Personal communication, Gus Koehler, Disaster Medical Response Planner, Emergency Medical Services Authority, State of California, Sacramento, 1987.

(Kreimer, 1980)
Kreimer A: The role of the mass media in disaster reporting: a search for relevant issues. In: Disasters and the mass media, Committee on Disasters and the Mass Media, Commission on Sociotechnical Systems, National Research Council—National Academy of Sciences, Washington, DC, 1980.

(Kreps, 1980)
Kreps GA: Research needs and policy issues on mass media disaster reporting. In: Disasters

and the mass media, Committee on Disasters and the Mass Media, Commission on Sociotechnical Systems, National Research Council—National Academy of Sciences, Washington, DC, 1980.

(Kusler, 1985)
Kusler JA: Liability as a dilemma for local managers, In Petak WJ: Emergency management: a challenge for public administration, special issue, Public Admin Rev 45:118, Jan 1985.

(Lane, 1984)
Lane HU, editor: The world almanac and book of facts—1985, New York, 1984, Newspaper Enterprise Association, Inc.

(Lantis, 1984)
Lantis M: When the earthquake hits home: Anchorage in the "Great Alaska Earthquake," Natural Hazards Research and Applications Information Center, University of Colorado, Boulder, 1984.

(Larson, 1980)
Larson JF: A review of the state of the art in mass media disaster reporting. In: Disasters and the mass media, Committee on Disasters and the Mass Media, Commission on Sociotechnical Systems, National Research Council—National Academy of Sciences, Washington, DC, 1980.

(Lavalla, 1983)
Lavalla R, and Stoffel S: Blueprint for community emergency management: a text for managing emergency operations, Emergency Response Institute, Olympia, WA, 1983.

(Lewis, 1980)
Lewis FR, Trunkey DD, and Steele MR: Autopsy of a disaster: the Martinez bus accident, J Trauma 20(10):861, 1980.

(Lowry, 1983)
Lowry S: Triage protocols and guidelines, adopted by the Sacramento County Emergency Medical Care Committee, August 9, 1983 (unpublished), Sacramento County, CA 1983.

(Marks, 1954)
Marks ES, et al: Human reactions in disaster situations, vols 1–3, unpublished report, National Opinion Research Center, University of Chicago, June 1954.

(Mattingly, 1988)
Mattingly S: Picture windows of opportunity, Nat Haz Observer 12(3):1, Jan 1988.

(Maxwell, 1982)
Maxwell C: Hospital organizational response to the nuclear accident at Three Mile Island: implications for future-oriented disaster planning. In Cowley RA, editor: Mass casualties: a lessons learned approach, p 295. Proceedings of the first international assembly on emergency medical services, Baltimore, MD, June 13–17 1982, DOT HS 806 302, US Department of Transportation, National Highway Traffic Safety Administration, Washington, DC, 1982.

(May, 1985)
May PJ: FEMA's role in emergency management: examining recent experience. In Petak WJ: Emergency management: a challenge for public administration (special issue), Public Admin Rev 45:40, Jan 1985.

(Mcluckie, 1970)
Mcluckie B: The warning system in disaster situations: a selective analysis, report series no 9, Disaster Research Center, University of Delaware, Newark, 1970.

(Mesnick, 1980)
Mesnick PS: Value of disaster critiques as demonstrated by the management of two "L" crashes in the City of Chicago. In Frey R, and Safar P: Types and events of disasters: organization in various disaster situations, New York, 1980, Springer-Verlag.

(Mileti, 1975)
Mileti DS, Drabek TE, and Haas JE: Human systems in extreme environments: a sociological perspective, monograph no 21, Program on Technology, Environment and Man, Institute of Behavioral Science, University of Colorado, Boulder, 1975.

(Moore, 1958)
Moore HE: Tornados over Texas: a study of Waco and San Angelo in disaster, Austin, 1958, University of Texas Press.

(Moore, 1964)
Moore HE: And the winds blew, Hogg Foundation for Mental Health, University of Texas, 1964.

(Moore, 1967)
Moore WS: A new classification system for disaster casualties, Hospitals—JAHA 41(4):66, Feb 16, 1967.

(Morris, 1981)
Morris GP: Preplan was the key to MGM rescue

response as EMS helped thousands of hotel fire victims, Fire Command 48:20, June 1981.

(Morris, 1982)
Morris GP: The Kenner airliner disaster: a 727 falls into a New Orleans suburb, J Emerg Med Serv 7(9):58, Sept 1982.

(Munninger, 1981)
Munninger K, and Ravenholt O: Lessons from Las Vegas: the MGM Hotel fire, J Emerg Med Serv 6(2):37, Feb 1981.

(Mushkatel, 1985)
Mushkatel AH, and Weschler LF: Emergency management and the intergovernmental system. In Petak WJ: Emergency mangement: a challenge for public administration, special issue, Public Admin Rev 45:49, Jan 1985.

(NAB)
Natural disasters: A broadcaster's guide, National Association of Broadcasters, Washington, DC, no date.

(NHRAIC, 1987)
San Francisco update, Nat Haz Observer 11(3):13, Jan 1987.

(NTSB, 1982)
National Transportation Safety Board: Railroad accident report: derailment of Washington Metropolitan Area Transit Authority train no 410 at Smithsonian interlocking, January 13, 1982, NTSB-RAR-82-6, National Transportation Safety Board, Bureau of Accident Investigation, Washington, DC, 1982.

(Nat Safety Council, 1973)
National Safety Council: Accident Facts, Chicago, IL, 1973.

(Neff, 1977)
Neff JL: Responsibility for the delivery of emergency medical services in a mass casualty situation: the problem of overlapping jurisdictions, Mass Emergencies 2:179, 1977.

(Orr, 1982)
Orr SM, and Robinson WA: The Hyatt disaster: two physicians' perspectives, J Emerg Nurs 8(1):6, Jan/Feb 1982.

(Orr, 1983)
Orr SM: The Hyatt Regency skywalk collapse: an EMS-based disaster response, Ann Emerg Med 12(10):601, 1983.

(Parr, 1970)
Parr AR: Organizational response to community crises and group emergence, Am Behav Sci 13(3):423, Jan–Feb 1970.

(Parrish, 1981)
Parrish RL: The MGM Grand Hotel fire: an overview of the fire fighting and evacuation operations, International Fire Chief 47:12, Jan 1981.

(Patterson, 1981)
Patterson P: OR staffs respond to Hyatt casualties, AORN J 34(3):411, Sept 1981.

(Perkins, 1984)
Perkins J, project manager: The liability of businesses and industries for earthquake hazards and losses, Earthquake Preparedness Program, Association of Bay Area Governments, Oakland, CA, 1984.

(Perry, 1985)
Perry RW: Comprehensive emergency management: evacuating threatened populations, Greenwich, CT, 1985, JAI Press, Inc.

(Petak, 1985)
Petak WJ: Emergency management: a challenge for public administration. In Petak WJ: Emergency management: a challenge for public administration, special issue, Public Admin Rev 45:3, Jan 1985.

(Publ Safety Dept, 1975)
Public Safety Department, Lakewood, Colorado: Ten code versus clear speech communication, Commission on Peace Officer Standards and Training, State of California, Sacramento, 1975.

(Pyle, 1946)
Pyle E: Last Chapter, New York, 1946, Henry Holt & Co.

(Quarantelli, 1960)
Quarantelli EL: Images of withdrawal behavior in disasters: some basic misconceptions, Soc Prob 8(1):68, Summer 1960.

(Quarantelli, 1965)
Quarantelli EL: Mass behavior and governmental breakdown in major disasters: viewpoint of a researcher, Police Yearbook 105–112, 1965.

(Quarantelli, 1970a)
Quarantelli EL: The community general hospital: its immediate problems in disasters, Am Behav Sci 13(3):389, Jan–Feb 1970.

(Quarantelli, 1970b)
Quarantelli EL, and Dynes RR: Introduction special issue on organizational and group behavior in disaster, Am Behav Sci 13(3):325, Jan–Feb 1970.

(Quarantelli, 1972)
Quarantelli EL, and Dynes RR: When disaster strikes (it isn't much like what you've heard or read about), Psychology Today, p 67, Feb 1972.

(Quarantelli, 1978)
Quarantelli EL: Some basic themes in sociological studies of disasters. In Quarantelli EL: Disasters: theory and research, Beverly Hills, CA, 1978, Sage Publications.

(Quarantelli, 1979a)
Quarantelli, EL: Studies in disaster response and planning, final project rep no 24, Disaster Research Center, University of Delaware, Newark, 1979.

(Quarantelli, 1979b)
Quarantelli EL, and Tierney KJ: Disaster preparation planning. In: Office of Public Sector Programs, Fire safety and disaster planning, 263-313, reprint no 118, American Association for the Advancement of Science, Washington, DC, Disaster Research Center, University of Delaware, Newark, 1979.

(Quarantelli, 1981a)
Quarantelli EL: Disaster planning: small and large—past, present and future. Presented at the American Red Cross EFO Division Disaster Conference, Blacksburg, VA, Feb 19–22, 1981, Disaster Research Center, University of Delaware, Newark, 1981.

(Quarantelli, 1981b)
Quarantelli EL: The command post point of view in local mass communications systems, Int J Comm Res 7:57, 1981, preliminary paper no 150, Disaster Research Center, University of Delaware, Newark, 1981.

(Quarantelli, 1981c)
Quarantelli EL: Sociobehavioral responses to chemical hazards: preparations for and responses to acute chemical emergencies at the local community level, Disaster Research Center, University of Delaware, Newark, 1981.

(Quarantelli, 1982a)
Quarantelli EL: Inventory of disaster field studies in the social and behavioral sciences 1919–1979, Disaster Research Center, University of Delaware, Newark, 1982.

(Quarantelli, 1982b)
Quarantelli EL: Human resources and organizational behaviors in community disasters and their relationship to planning, preliminary paper no 76, Disaster Research Center, University of Delaware, Newark, 1982.

(Quarantelli, 1982c)
Quarantelli EL: The Grand Island, Nebraska, tornado case study: emergency sheltering aspects. In: Sheltering and housing after major community disasters: case studies and general observations, p 55, Disaster Research Center, University of Delaware, Newark, 1982.

(Quarantelli, 1983)
Quarantelli EL: Delivery of emergency medical care in disasters: assumptions and realities, New York, 1983, Irvington Publishers, Inc.

(Quarantelli, 1985)
Quarantelli EL: Organizational behavior in disasters and implications for disaster planning, report series 18, Disaster Research Center, University of Delaware, Newark, 1985.

(Quarantelli, 1987)
Quarantelli EL: Personal communication, Co-director, Disaster Research Center, University of Delaware, Newark, Jan 13, 1987.

(Raether, 1986)
Raether HC: Legal and financial considerations in multiple death disasters. In: Multiple death disaster response workshop, SM/IG 193, Federal Emergency Management Agency and the National Funeral Directors Association, Washington, DC, 1986.

(Raker, 1956)
Raker JW, Wallace AFC, and Rayner JF: Emergency medical care in disasters: a summary of recorded experience, study no 6, publication no 457, Disaster Research Group, National Academy of Sciences—National Research Council, Washington, DC, 1956.

(Reynolds, 1976)
Reynolds S, and Wright JE: A selective literature review of disaster medical services, working paper no 64, Disaster Research Center, University of Delaware, Newark, 1976.

(Ringhofer)
Ringhofer, JC: Technology transfer: law en-

forcement application of the incident command system, California Commission on Peace Officer Training, Center for Executive Development, Command College, Class No. 1, Core III, Independent Study Project, San Bernardino County Sheriff's Department, San Bernardino, CA, no date.

(Ritz, 1980)
Ritz, WR: A case study of newspaper disaster coverage: the Big Thompson Canyon Flood. In: Disasters and the mass media, Committee on Disasters and the Mass Media, Commission on Sociotechnical Systems, National Research Council—National Academy of Sciences, Washington, DC, 1980.

(Rosow, 1977)
Rosow I: Authority in emergencies: four tornado communities in 1953, The Disaster Research Center historical and comparative series, no 2, Disaster Research Center, University of Delaware, Newark, 1977.

(Ross, 1982)
Ross RB: Emergency planning paid off, Security Management, p 62, Sept 1982.

(Roth, 1970)
Roth R: Cross-cultural perspectives on disaster response, Am Behav Sci 13(3):440, Jan–Feb 1970.

(Rubin, 1987)
Rubin DM: Lessons learned from Three Mile Island and Chernobyl: how the news media report serious nuclear power plant accidents, Emergency Prepared Digest (Canada) 14(2):10, April–June 1987.

(Ruch, 1984)
Ruch C: Human response to vertical shelters: "An experimental note," Int J Mass Emerg Dis 2(3):389, Nov 1984.

(Rund, 1981)
Rund DA, and Rausch TS: Triage, St Louis, 1981, The C.V. Mosby Co.

(Sac Fire Com, 1986)
Fire Communication Center: Report of regional disaster medical plan evaluation, unpublished, Sacramento, CA, 1986.

(San Bernardino County, 1985)
San Bernardino County Sheriff's Department: Law enforcement incident command field operations guide, San Bernardino, CA, 1985.

(Savage, 1977)
Savage P: Hospital disaster planning. In Easton K, editor: Rescue emergency care, London, 1977, William Heinemann Medical Books, Ltd.

(Scanlon, 1980)
Scanlon J: The media and the 1978 Terrace Floods: an initial test of a hypothesis. In: Disasters and the mass media, Committee on Disasters and the Mass Media, Commission on Sociotechnical Systems, National Research Council—National Academy of Sciences, Washington, DC, 1980.

(Scanlon, 1982)
Scanlon TJ, and Alldred S: Media coverage of disasters: the same old story, Emergency Planning Digest (Canada) 7(4):13, 1982.

(Scanlon, 1985)
Scanlon J, et al: Coping with the media in disasters: some predictable problems. In Petak WJ: Emergency management: a challenge for public administration, special issue, Public Admin Rev 45:123, Jan 1985.

(Scanlon, 1988)
Scanlon J: Unpublished data on the Edmonton, Ontario, Canada, tornado of July 31, 1987, obtained from Russell Dynes, University of Delaware, Disaster Research Center, Newark, 1988.

(Scherr, 1988)
Scherr M: Personal communication, Deputy Chief, Fire & Rescue Division, State of California, Office of Emergency Services, Executive Director, FIRESCOPE, Operations Coordination Center, Riverside, CA, 1988.

(Scholl, 1984)
Scholl RE, and Stratta JL, editors: Coalinga, California, earthquake of May 2, 1983: reconnaissance report, Earthquake Engineering Institute, Berkeley, 1984.

(Schumacher, 1988)
Schumacher J: Sharing a training facility, Fire Command 55(3):16, March 1988.

(Seismic Safety Comm, 1979)
Seismic Safety Commission: Public official attitudes toward disaster preparedness in California, publication no SSC 79-05, Sacramento, 1979.

(Seismic Safety Comm, 1983)
Seismic Safety Commission: Preliminary reports submitted to the Seismic Safety Commis-

sion on the May 2, 1983, Coalinga earthquake, publication no SSC 83-08, Sacramento 1983.

(Shaftan, 1962)
Shaftan GW: Disaster and medical care, J Trauma 2:111, 1962.

(Silverstein, 1984)
Silverstein ME: Triage decision trees and triage protocols: changing strategies for medical rescue in civilian mass casualty situations, US Department of Commerce, National Technical Information Service, Springfield, VA, 1984.

(Sorensen, 1981)
Sorensen JH: Emergency response to Mount St Helens' eruption: March 20 to April 10, 1980, working paper series no 43, Natural Hazards Research and Applications Information Center, University of Colorado, Boulder, 1981.

(Sorensen, 1985)
Sorensen JH, Mileti DS, and Copenhaver E: Inter- and intraorganizational cohesion in emergencies, Mass Emergencies and Disasters 3(3):27, Nov 1985.

(Spirgi, 1979)
Spirgi EH: Disaster management: comprehensive guidelines for disaster relief, Bern, 1979, Hans Huber Publishers.

(Stallings, 1971)
Stallings R: Communications in natural disasters, rep no 10, Disaster Research Center, University of Delaware, Newark, 1971.

(Starr, 1969)
Starr C: Social benefit versus technological risk: what is our society willing to pay for safety?, Science 165:1232, Sept 19, 1969.

(State of California, 1982)
State of California, Office of Emergency Services: Field operations guide, ICS 420-1, Fire and Rescue Division, Riverside, 1982.

(Stevenson, 1981)
Stevenson L, and Hayman M: Local government disaster protection: final technical report, International City Management Association, Washington, DC, 1981.

(Stout, 1981)
Stout J, and Smith P: Nightmare in Kansas City, J Emerg Med Serv 6(9):32, Sept 1981.

(Super, 1984)
Super G: START: A triage training module, Hoag Memorial Hospital Presbyterian, Newport Beach, CA, 1984.

(Swint, 1976)
Swint C: Accident, ammonia, test hospital's disaster plan, Mod Health Care 6:44, Aug 1976.

(Thomas, 1988)
Thomas J: Washington report: Becton cites civil defense as top FEMA priority, Fire Chief 32(2/whole no 370):14, Feb 1988.

(Tierney, 1977)
Tierney KJ, and Taylor VA: EMS delivery in mass emergencies: preliminary research findings, Mass Emergencies 2:151, 1977.

(Tierney, 1980)
Tierney KJ: A primer for preparedness for acute chemical emergencies, book and monograph series no 14, The Disaster Research Center, University of Delaware, Newark, 1980.

(Tierney, 1985a)
Tierney KJ: Emergency medical preparedness and response in disasters: the need for interorganizational coordination. In Petak WJ: Emergency management: a challenge for public administration, special issue, Public Admin Rev 45:77, Jan 1985.

(Tierney, 1985b)
Tierney KJ: Report on the Coalinga earthquake of May 2, 1983, publication no SSC 85-01, Seismic Safety Commission, State of California, Sacramento, 1985.

(Tintinalli, 1978)
Tintinalli J, editor: A study guide in emergency medicine, American College of Emergency Physicians, Dallas, TX, 1978.

(Troeger, 1976)
Troeger J: The Xenia disaster, Emerg Product News, p 126, April 1976.

(US Dept Defense, 1975)
United States Department of Defense: Emergency war surgery: first United States revision of the emergency war surgery NATO handbook, US Government Printing Office, Washington, DC, 1975.

(US Fire Admin, 1980)
United States Fire Administration: The national workshop for fire services EMS needs: the Rockville report, Federal Emergency Management Agency, Washington, DC, Feb 1980.

(Vines, 1986)
Vines T: What future MSS [Mobile Satellite Service]? Clogging the prospects for more efficient communications?, Response, J Nat Search and Rescue Assoc 5(3):10, May/June 1986.

(Walker, 1982)
Walker B: Planet earth: earthquake, Alexandria, VA, 1982, Time-Life Books.

(Wallace, 1956)
Wallace AFC: An exploratory study of individual and community behavior in an extreme situation: tornado in Worcester, study no 3, publication no 392, Committee on Disaster Studies, National Academy of Sciences—National Research Council, Washington, DC, 1956.

(Wallace, 1985)
Wallace WA, and De Balogh F: Decision support systems for disaster management. In Petak WJ: Emergency management: a challenge for public administration, special issue, Public Admin Rev 45:134, Jan 1985.

(Wenger, 1975)
Wenger DE, Dykes JD, and Sebok TD: It's a matter of myths: an empirical examination of individual insight into disaster response, Mass Emergencies 1(1):33, Oct 1975.

(Wenger, 1978)
Wenger DE: Community response to disaster: functional and structural alterations. In Quarantelli EL: Disasters: theory and research, Beverly Hills, CA, 1978, Sage Publications.

(Wenger, 1980)
Wenger DE: A few empirical observations concerning the relationship between the mass media and disaster knowledge: a research report. In: Disasters and the mass media, Committee on Disasters and the Mass Media, Commission on Sociotechnical Systems, National Research Council—National Academy of Sciences, Washington, DC, 1980.

(Wenger, 1985a)
Wenger DE, James TF, and Faupel CE: Disaster beliefs and emergency planning, New York, 1985, Irvington Publishers, Inc.

(Wenger, 1985b)
Wenger D: Mass media and disasters, preliminary paper no 98, Disaster Research Center, University of Delaware, Newark, 1985.

(Wenger, 1986)
Wenger D, Quarantelli EL, and Dynes R: Disaster analysis: emergency management offices and arrangements, final project report no 34, Disaster Research Center, University of Delaware, Newark, 1986.

(White, 1981)
White D: In-depth report reveals new evidence: MGM update, Firehouse 6:26–32,59,71, Feb 1981.

(Williams, 1956)
Williams HB, and Rayner JF: Emergency medical services in disaster, Med Ann District of Columbia 25(12):655, Dec 1956.

(Wohlwerth, 1987)
Wohlwerth N: Putting computer technology to work in emergency planning, Emerg Prepared Digest (Canada) 14(2):6, April–June 1987.

(Worth, 1977)
Worth MF, and Stroup J: Some observations of the effect of EMS law on disaster related delivery systems, Mass Emergencies 2:159, 1977.

(Wright, 1976)
Wright JE: Interorganizational systems and networks in mass casualty situations, dissertation, The Ohio State University, 1976.

(Wright, 1977)
Wright JE: The prevalence and effectiveness of centralized medical responses to mass casualty disasters, Mass Emergencies, 2:189, 1977.

(Yutzy, 1969)
Yutzy D: Community priorities in the Anchorage, Alaska, earthquake, 1964, Disaster Research Center monograph series, no 4, The Disaster Research Center, University of Delaware, Newark, 1969.

(Zimmerman, 1985)
Zimmerman R: The relationship of emergency management to governmental policies on man-made technological disasters. In Petak WJ: Emergency management: a challenge for public administration, special issue, Public Admin Rev 45:29, Jan 1985.

(1971)
The day the earth shook, Emerg Med, p 27, April 1971.

(1975)
Moorgate Tube train disaster, Brit Med J, p 727, Sept 27, 1975.

(1977)
Trends in triaging: A random survey of triage tags used in the United States, Emerg Prod News, p 88, Oct 1977.

(1982)
Media rapped and lauded for coverage of US disasters, Emerg Dept News 4(3):19, March 1982.

(1983)
Incident command system, ed 1, Fire Protection Publications, Oklahoma State University, Stillwater, OK, 1983.

INDEX

Auf der Heide, *Disaster Response*

(Page numbers in *italics* indicate figures. Page numbers followed by a *t* indicate tables.)

APPLICATION FOR PHYSICIAN CATEGORY I

CONTINUING MEDICAL EDUCATION CREDIT

Disaster Response: Principles of Preparation and Coordination has been approved by the American College of Emergency Physicians for 25 hours of continuing medical education credit. This is available upon payment of an application fee and satisfactory completion of a written exam. This project is co-sponsored by the California Chapter of the American College of Emergency Physicians.

Applicant's Name: _____
                        Last   First   Middle Initial

Degree (MD, DO) _____   Phone (   )_____ - _____

Address: _____
        Institution

_____
Number   Street

_____
City   State   Country   Zip Code

Application Date: _____
                Mo   Day   Year

Send with non-refundable application fee of $120.00* to:

        Disaster Response Continuing Medical Education
        Erik Auf der Heide, M.D.
        c/o California Chapter of ACEP
        505 N. Sepulveda Blvd., Suite 12
        Manhattan Beach, CA 90266

(Make checks payable to Erik Auf der Heide, MD)

Applicants will receive exam (each question indicates the text page numbers on which the answers can be found), answer sheet, and statement of text learning objectives. Allow 6 weeks after returning exam to the address above for certificate delivery.

*Application fee does not include cost of the textbook and is subject to change without notice.